БОРАТНЫЕ СТЕКЛА

BORATNYE STEKLA

BORATE GLASSES

BORATE GLASSES

Thermochemical Processes in Glass Formation, Crystallo-optics, Technology, Physicochemical Properties, and Structure of Glasses with the Composition B_2O_3–Li_2O–MeO

L. Ya. MAZELEV

TRANSLATED FROM RUSSIAN

Springer Science+Business Media, LLC

1960

Original Russian text published by the Academy of Sciences
Belorussian SSR Press in Minsk, 1958

Editor: L.K. Petrov

Library of Congress Catalog Card Number 60-8719

ISBN 978-1-4899-4679-9   ISBN 978-1-4899-4677-5 (eBook)
DOI 10.1007/978-1-4899-4677-5

TABLE OF CONTENTS

INTRODUCTION

The science and technique of glass manufacture are at a very high level at the present time. A number of complicated problems in the chemistry and technology of glass manufacture, thermotechnics, and the mechanization and automation undertaken by the glass industry have been solved successfully; thousands of qualified engineers, experts,and scientific workers have been trained. There are a number of large specialized scientific organizations as well as university departments of silicate technology. However, up to now the problem of the structure of glass, especially that of borate glass, has not been solved conclusively. The term "glass", too, has as yet no generally accepted, unequivocal definition [1-18].

At the All-Union Conference on the Structure of Glass, A. A. Lebedev [13] noted that as a result of the conference a general point of view on the structure of glass had been established, namely, that glass has an amorphous crystallite structure. The problem of the volume and role of the ordered sections in glass and their effect on glass properties, and of the changes in glass properties with changes in composition, etc., has remained unsolved. The conference recognized the need for further study of glasses in order to establish a well-grounded theory concerning their structure.

This book is based mainly on the results of investigating silicate and, to a very limited extent, borate glasses. However, a mechanical extension of the rules for one type of glass to cover another type (in this case, borate glass) would be a bad error.

Little is known about the role and behavior of B_2O_3 in glasses, the effect of this oxide on their properties, or the structure of borate glasses.

Boron as an element has been known for a long time but the role of its oxide in glass has been studied little, and literature data are often contradictory [4, 39, 89].

The literature gives the results of x-ray investigations of H_3BO_3. Zachariasen and others [43-46] consider that the main structural element of crystalline boric acid is a flat, almost equilateral triangle of $(BO_3)^{3-}$. Such triangles are joined by hydrogen bonds to give a hexagon with two boron atoms and four oxygen atoms. Adjacent oxygen atoms are bound by hydrogen atoms.

On the basis of measurements of the heat capacity of glasses at low temperatures, V. V. Tarasov [48] considered that Zachariasen's point of view on the two-dimensional polymerization of B_2O_3 was incorrect. Like a series of others, borate glasses have a chain structure. The boron atoms form the apexes of low, three-faced pyramids of $(BO_3)^{3-}$ with the apexes of adjacent pyramids arranged alternately on one side or the other of the oxygen atoms. According to Tarasov, every glass has a three-dimensional branched skeleton. Richter and others [47] hold similar points of view.

The structure and properties of natural borates have been studied comparatively fully [39, 45], but this cannot be said of borates in a glassy state. The $B_2O_3 - SiO_2$ system was partially studied in 1928 by Cousen and Turner [52], $MgO - B_2O_3$ by Toropov and Konovalov in 1940 [52, 53] and Davis and Knight in 1945 [54, 55], $CaO - B_2O_3$ was partially studied in 1932 by Carlson [52, 55] and $PbO - B_2O_3$ by Bunting and Geller in 1937 [52, 55]. The $B_2O_3 - Li_2O$ system was studied thoroughly with respect to glass formation by Mazetti and de Carli in 1926 [55, 56] and with respect to physicochemical properties and structure by Bresker and Evstrop'ev [58, 59] in 1952.

Among the ternary boron systems, the following have been studied: $CaO - B_2O_3 - SiO_2$ by Flint and Wells in 1936 [52]; $PbO - B_2O_3 - SiO_2$ by Geller and Bunting in 1939 [52], and $B_2O_3 - BeO - Na_2O$ by Menzel and Adam in 1949 [70, 76, 77]. The $B_2O_3 - BeO - Li_2O$ system was studied by Lindemann in 1911 [78], Schleede and Wellmann in 1913 [76], Botvinkin in 1931 [79], Gertsriken and Tanchakivskii in 1936 [80, 81], and fully for the first time by Mazelev in 1940 [52, 61] and then by Menzel and Slivinski in 1942 [75]. The $B_2O_3 - Li_2O - MgO$ system was studied by Mazelev in 1953 [62].

Moulton [82] patented a glass with the composition $B_2O_3 - CdO - BeO$; Sun and Callear [83], a cadmium lanthanum borate; and Armistead [84], a cadmium borosilicate. Lanthanum cadmium fluoborates, Li, Zn, Ba, Pr, Th, Sn, and Ta cadmium borates [86], and cadmium borosilicates, titanates, zirconates, etc., [87] have been studied more or less thoroughly as thermal-neutron-absorbing glasses.

Among the quaternary boron (nonsilicate) systems, only the $B_2O_3 - Li_2O - BeO - MgO$ system has been studied systematically (Mazelev, 1952 [63, 64]).

This is approximately the extent to which borates have been studied and it throws very little light on either the problem of the structure of borate glasses or the character of the effect of B_2O_3 on glass properties.

The literature contains some data on the effect of B_2O_3 on the properties of silicate glasses and enamels [4, 39, 89]. There are no data on the effect of B_2O_3 on borate (nonsilicate) glasses. Contemporary practice deals with glasses in which B_2O_3 is the predominating component and the only acidic oxide. It is therefore necessary to study the conditions and limits of application of B_2O_3 in glasses and also analyze occasionally the contradictory literature data on this problem.

An analysis of the property curves of borate glasses shows a series of anomalies. Bresker and Evstrop'ev [59] found unusual points on the composition − property curves and concluded that in boroalkali glasses containing up to 16 mol. % of alkaline oxide the boron atoms changed their coordination number from three to four (see Table 2 in appendix for the compositions of glasses of the $B_2O_3 - Li_2O$ system). On analyzing the results of investigations of binary boro-alkali systems, they concluded that with the small radius of the Li^+ cation the property curves obeyed a linear relation, but as the dimensions of the cation increased in passing from Li^+ to Na^+ and K^+, the character of the curves changed sharply. The authors considered that Li^+ evidently filled spaces in the borate glass structure without distorting the structure itself. The spaces, however, were difficultly accessible to the larger ions Na^+ and K^+, at least without structural changes. Thus, the degree and character of the changes in the properties of borates to which alkaline oxides were added were determined by the radius of the alkali cations, while the property changes of glasses with divalent oxides added were determined by the charge of the cation. In this case, Evstrop'ev's opinion [22] that the properties of glass are related to the phase diagram is confirmed.

Bresker and Evstrop'ev also considered the role of an oxygen ion in the formation of the glass structure [59]. According to them, it is precisely the oxygen atoms which play the main role in the formation of the glass structure and not the tendency of boron to change its coordination number from three to four in relation to oxygen. Therefore, the course of the composition − property curves in binary boroalkali glasses is related to the changing values of the

2

volume of the oxygen ion in the glass. The authors also concluded that definite chemical compounds, detected in the phase diagrams of the given system, exist in glasses.

According to Ainsworth [130], the maximum effect of the anomaly of the properties in the $B_2O_3 - Na_2O$ system is observed at an Na_2O content of 16.7 mol. % and is related to the conversion of 1/5 of the total amount of boron from BO_3 to BO_4. Excess Na_2O did not result in an increase in the number of BO_4 tetrahedra, but passed into the lattice spaces and promoted rupture of the oxygen bonds.

According to Warren and Biscoe [90], in these alkali borates the physical properties of the glasses change until the ratio O/B = 1.61. Stevels [73] elaborated this idea and divided the zone of borate formation into two: an accumulation zone where each oxygen atom is bound to two boron atoms and a structural disruption zone where some tetrahedra have oxygen with a single bond. The boundary of these two zones is the molecular ratio O/B = 1.61.

Laurent [73] reported that in the interval between O/B = 1.71-1.78 the coefficient of expansion hardly changes, although there is an increase in the amount of unsaturated oxygen. Between O/B = 1.78-1.93, the increase in the unsaturated oxygen content, if caused by an increase in alkali, i.e., mobile ions, would produce lattice deformation that was hardly noticeable for lithium glasses (Li^+ ion) but increased from Li^+ to K^+. Thus, the specific consolidating effect of the Li^+ ion with its high field strength on the lattice becomes noticeable.

Facts on the anomaly in the properties of borate glasses were also established by other investigators. Shartsis et al. [91] established that over the temperature range of 600-1000° there were maxima and minima for a number of physical properties of glasses with a continuous change in the B_2O_3 content; this is also explained by a change in the coordination number of part of the boron. An anomalous change in viscosity was noted for borate glasses in the range of temperatures corresponding to glass viscosities of the order of 10^{11}-10^{12} poises [92].

English [74], King and Andrews [94], Gehlhoff and Thomas [95], and Jamamoto and Kishii [96] explain the unusually low value of the coefficient of expansion of B_2O_3 (below the "transformation point") by the formation of BO_4 tetrahedra at the given temperatures. They explained the strong fluxing action of B_2O_3 in the melting of glass by the formation of flat triangles of BO_3. Gooding and Turner [99] also confirmed the presence of these anomalies on determining the coefficient of expansion.

Indenbom [103] observed in glasses of the "Nonex" and No. 46 types an anomalous birefringence, corresponding to strains of the first and second order. The removal of these strains required special glass-annealing conditions.

Botvinkin presents a series of expansion curves for glass samples with a gradually changing composition from silica, through intermediate mixtures to pure boric oxide [20]. On going to high-boron glasses, the expansion curves continuously become steeper, indicating quite a rapid growth in the coefficient of expansion with an increase in the B_2O_3 content of the glass.

According to Botvinkin, the glasses expand almost linearly above and below the anomalous range. A transition region is observed in this anomalous range.

In developing their ideas on the structure of B_2O_3 and borate glasses, a number of authors have put forward a hypothesis concerning the formation of XY_4 groups in borate glasses [92]. The rule of XY_4 group formation states (X is BO_4, Y is BO_3):

1. One (each) X is attached to four Y's.

2. One (each) Y is attached to two Y's and one X.

3

3. The XY_4 groups are attached by the apexes but not by edges or faces. This means that the X groups are not directly attached to each other and that each Y is not attached to more than one X; one BO_4 tetrahedron is bound to four BO_3 groups, each of which is attached to the tetrahedron by the apex. The positive cations Na^+ are disposed close to X.

If the amount of Na_2O is less than that at which maximum BO_4 formation occurs, then BO_3 triangles are found simultaneously with XY_4 groups. If SiO_2 is added to the $Na_2O - B_2O_3$ system, then $[SiO_4]^{4-}$ tetrahedra also appear in this atomic group.

However, it should be considered that XY_4 groups are formed only over the definite temperature range between the annealing temperature and the softening point, which corresponds to a viscosity of the order of 10^{11} and 10^{12} poises. XY_4 groups are not formed at a lower temperature and at higher temperatures they change into form Y.

As already stated, in glasses of the ternary system $Na_2O - B_2O_3 - SiO_2$, not only XY_4 atomic groups are formed, but also XY_3Z, XY_2Z_2, XYZ_3 and XZ_4 groups. Z corresponds to the $[SiO_4]^{4-}$ tetrahedron, which replaces from one to four Y's. However, the tendency for these glasses to separate into two phases hinders the formation of groups different from XY_4. In this way these authors explain a number of anomalous physical properties of the glass and also a series of technological properties, in particular, the phenomena of opalization and devitrification and the relation to chemical stability, etc.

They consider that the phenomenon of birefringence reported by Indenbom is also explained by the formation of XY_4 groups and the separation of the glass under certain conditions into 2 phases: siliceous (more refractory) and borosodium (less refractory).

Appen [33-35, 105] also explained the anomalies in the properties of B_2O_3 by transition of part of the BO_3 and BO_4 at a definite ratio of $\dfrac{Na_2O}{B_2O_3}$. At a ratio of $\dfrac{Na_2O}{B_2O_3} < \dfrac{1}{3}$, the values of $N_{B_2O_3}$ and $V_{B_2O_3}$ change little and remain close to the properties of pure boric oxide. At a ratio of $\dfrac{Na_2O}{B_2O_3} > \dfrac{1}{3}$, the properties of the glass begin to change sharply, reaching limiting values ($N_{B_2O_3}$ = 1.710 and $V_{B_2O_3}$ = 18.5 cm^3/mole) at a definite alkali content and then remain approximately constant.

This is a short account of the opinions of those investigators who consider that all the anomalies and extrema in the changes in properties are explained by conversion of part of the BO_3 into BO_4 at a definite, quantitative content by Me_2O or the oxides MeO and also by the corresponding formation (according to Warren, Stevels, Laurent, Abe, and others) of XY_4 and $XY_nZ_{(4-n)}$ groups.

In their conclusions, Fajans and Barber [47] did not start from hypotheses on the effect of the coordination change of boron on the character of the changes in properties of borate and borosilicate glasses.

Not all investigators support the theory of glass structure presented by Warren et al. and Tsint and Moravets, Berger, Grjotheim, and Krogh-moe and others [107], also referring to the results of x-ray investigation of B_2O_3 and experimental determinations of its properties, proposed a directly opposite theory of the change in structure and properties of B_2O_3 with a change in composition and temperature. The authors showed that crystalline boric oxide consists mainly of BO_4 or, more accurately, is a mixture of BO_3 and BO_4 with the latter predominating. This was confirmed by Belov [13]. The boron atoms in the tetrahedra are arranged at a separation of 0.44 A. Such weak bonds with tetracoordination facilitate disintegration on melting. The BO_3 triangles are much stronger.

An increase in the alkali oxide content strengthens the structure of the glass; with the addition of a definite amount of alkali oxides, there is a rapid increase in the viscosity, softening point, density, and refractive index. The authors showed that this could not be explained by the change of BO_3 into BO_4 because less energy is required for rupture of the bonds in the BO_4 molecule, with its large interatomic distances (2.14 A), than in the BO_3 molecule, which has shorter bonds (1.48 A). This may be explained only by the conversion of BO_4 into BO_3 and not the reverse, as stated by Warren et al., [92].

This is a short account of the very incomplete and partly contradictory literature data on the role of B_2O_3 in glass formation and in the structure of boroalkali and borosilicate glasses. The problems of the state and the role of B_2O_3 in glasses where it is the sole acid oxide and the predominating component have not been solved.

It was quite correctly indicated in the resolutions of the Conference on the Structure of Glass [13] that there is a need for more precise determination of the dimensions and forms of the ordered regions in glasses, more complete investigation of the kinetics and mechanism of the processes in the change from a highly viscous to a glassy state, etc. All this should be achieved on the basis of data from experimental investigations of different systems.

We will consider briefly the properties of the other oxides which are components of the borolithium glasses investigated.

Lithium oxide is one of the most important components in the production of optical glasses, in particular those for optical apparatus [108, 109], electrovacuum glasses, and enamels, especially enamels for coating low-melting, nonferrous metals. A short review of lithium compounds and their field of application in glass has been presented in the literature [108-115].

The introduction of Li_2O into glass produces a decrease in the amount of alkali required, a fall in the thermal expansion coefficients of the glass, an increase in the density of glasses, a decrease in the tendency of glasses to crystallize, an increase in the refractive index of the glass, a rise in the dielectric properties of the glass, an increase in the transparency of the glass to ultraviolet radiation, etc.

In this case, also, contradictory data are encountered on the effect of Li_2O on the properties of glass: the crystallization capacity, the transparency to ultraviolet radiation, the thermal expansion, the chemical stability, etc. [76, 112, 117, 118].

As yet, beryllium oxide has limited application in glass production due to its shortage. BeO plays a very important part in atomic technology [119] for the preparation of reflectors and moderators of neutrons in reactors. The efficiency of BeO in this respect exceeds that of graphite by a factor of 2, heavy water by a factor of 11, and water by a factor of 67. Be is 17 times more transparent to x-rays than Al.

In glass manufacture, BeO is used in the production of fluorescent lamps [119], for special compositions of optical glass, in particular, for modern optical apparatus [119], for the production of glasses transparent to x-rays [61], etc.

Heyne [118] used beryllium oxide to prepare various series of fluoroberyllium glasses. These glasses had low chemical stability and a hardness of 2 on the Mohs scale. The softening points were from 160-300° and the expansion coefficients over the range 20-100° were from 216 to $325 \cdot 10^{-7}$. The glasses had a high transparency to ultraviolet (up to 2200 A) and

5

infrared radiation (up to 5.0-5.5 μ). The properties of analogous glasses were also presented by Karsten [121].

Chi Fang Lai and Silvermann presented a description of siliceous beryllosodium and potassium glasses [122-123], which were transparent to ultraviolet radiation. Mauri considered that these glasses had a low transparency to ultraviolet radiation [124].

Data on the effect of BeO on the properties of glasses were presented by Chi Fang Lai and Silvermann [122, 123], Becker [125], Gottfried [126], White, Shremp, Brecken, and Jor [127], Ziegler and Wellman [76] and others [127-129]. All agree that BeO imparts to glass a series of valuable physical and physicochemical properties. However, these data are frequently contradictory; also they do not give the compositions of all the glasses investigated, which makes it impossible to compare them There are hardly any data on the effect of BeO on the structure and properties of pure borate glasses.

The properties of the other components, CaO, ZnO, BaO, and PbO, have been studied in detail for silicate glasses and very little for borate glasses. The properties of the oxides CdO and SrO have been studied very little as the latter were of limited availability for glass production. Recently CdO has found wide application as an important component in the production of neutron-absorbing glasses [82-86].

Borolithium glasses have wide application in the most varied fields of industry and science and even greater prospects as electrovacuum, optical, x-ray transparent or x-ray (neutron) absorbing and reactor glasses, glasses with a high microhardness, glasses for the production of various glass fibers, enamels, etc.

The purpose of the present investigation was a systematic study of the technology of glass melting, the processes and reactions of glass formation, the crystallization of glasses and the composition of crystallization products, physicochemical properties and their relation to the composition and structure of glass and methods of calculating these properties.

GLASS FORMATION DIAGRAMS OF THE SYSTEMS
$B_2O_3 - Li_2O - MeO$ (BeO, MgO, CaO, ZnO, SrO, CdO, BaO and PbO); TECHNOLOGY AND CONDITIONS OF MELTING AND PROCESSING THE GLASSES OF THESE SYSTEMS

By carrying out a considerable number of exploratory and then more than 1500 main melts, we determined the sections of glass formation and crystallization on cooling the melts of the following systems:

Systems	Melting points of mixtures, °C
$B_2O_3 - Li_2O - BeO$	1000-1200
$B_2O_3 - Li_2O - MgO$	1000-1100
$B_2O_3 - Li_2O - CaO$	1000-1100
$B_2O_3 - Li_2O - ZnO$	1000-1200
$B_2O_3 - Li_2O - SrO$	1100-1200
$B_2O_3 - Li_2O - CdO$	1000-1100
$B_2O_3 - Li_2O - BaO$	1000-1100
$B_2O_3 - Li_2O - PbO$	1050-1100
$B_2O_3 - BeO - MgO$	up to 1300
$B_2O_3 - Li_2O - BeO - MgO$	1000-1300
$B_2O_3 - Li_2O - BeO - MgO$ + from 1 to 70 parts by weight of SiO_2, Al_2O_3, or ZrO_2	1100-1400
$B_2O_3 - Li_2O - ZnO(CdO, SrO)$ + from 1 to 50 parts by weight of SiO_2, Al_2O_3, ZrO_2, or TiO_2	up to 1350.

Chemically pure materials were used for the charges. All the charges were first sintered at temperatures of 500-650° (maximum), i.e., till H_2O vapor was completely removed from the charges. The temperature was raised extremely slowly to avoid swelling of the charge. The sintered charge was then powdered and loaded into a melting crucible. The charge was melted with the most rapid rise in temperature possible. The melting points of the glasses were simultaneously the clearing and homogenization temperatures of the glass mixture due to the very low viscosity of the melt. Prolonged melting of the charge and also holding the melt in the furnace led to rapid volatilization of the components of the glass mixture, especially B_2O_3. The results of the main melts are presented in Figs. 1-8. It is not worthwhile to present the compositions of all the glasses. The limits of the amounts of components in glasses which did not crystallize when the melts were cooled are presented in Table 1.

The upper annealing temperature of the glasses, established from their softening points, was within the range 450-600°. However, annealing at the upper temperature limit facilitated crystallization of the glasses and the production of liquation in some of them, in particular, in borolithioberyllium and beryllomagnesium glasses. In these glasses the softening points practically coincided with the temperatures at which crystallization began. These glasses were therefore annealed at lower temperatures and for a longer time, i.e., at temperatures of not more than 400-500°.

Analysis of the results and melting points of these glass compositions shows that in the given systems where only the oxides MeO are different but the molecular-percent content of them in all the systems are the same, the tendency toward glass formation and crystallization is determined to a considerable extent by the nature of the oxide MeO and the character of its bond with B_2O_3. The greater the molecular weight of the oxides MeO and the value of their interatomic separations, then the greater is the tendency of the borates of MeO toward glass formation, and the greater the relative fusibility of the glass of the corresponding ternary system.

The results of studying the processes and reactions of glass formation showed that in the systems investigated, with an increase in temperature, vitrification began, as a rule, due to the formation of eutectics and reactions between the solid and liquid phases or between liquid phases of different compositions, not just by mechanical solution of the more refractory oxides in the melt of the more fusible ones.

The conservation of the glassy state with a reduction in temperature was achieved by a decrease in the kinetic energy of the component parts of the melt, by an increase in the intermolecular (interatomic) interaction and density of the particles due to a considerable increase in the viscosity, and also by the nature of the borates formed in the glass mixture and the degree of cohesion of B_2O_3 in the borates.

The greatest tendency to glass formation was shown by lithium borates with an Li_2O content in the glass of up to 15 wt. (25 mol.) % and this is connected with the effect of this amount of Li_2O on the change in the coordination state of the boron and on the change in the $BO_3 : BO_4$ ratio. An Li_2O content of more than 15 wt. (25 mol.) % in beryllium glasses promotes the production of liquation in the glass mixture on cooling or on subsequent heat treatment. We made use of this phenomenon to extract beryllium borate with the composition $3BeO \cdot B_2O_3$ from the glass mixture by treating it with dilute HCl solution. This also confirmed the complexity of the structure of these glasses and, to a certain extent, the role of Be^{2+} in the formation of this structure.

The effect of MgO on glass formation was similar to Li_2O. The simultaneous introduction of BeO + MgO into the composition of the glass considerably extended the glass-formation region and made it possible to introduce up to 22.5-25 wt. % of BeO + MgO into the glass. Here the glassy state was more stable when the content of MgO>BeO, i.e., when magnesium borates predominated.

Analysis of the results of melting glasses with the composition $Li_2O - BeO - MgO$, i.e., containing BeO and MgO simultaneously, shows that with a 1% MgO content, the glasses behaved like pure beryllium glasses on melting and cooling, i.e., 1% MgO had no effect on their behavior on cooling; glasses with $\dfrac{Li_2O}{BeO + MgO} \leq 1$ as a rule crystallized either completely or partially on cooling. With an MgO content of 8%, glasses did not crystallize on cooling even when they contained only 5% Li_2O, i.e., at a ratio $\dfrac{Li_2O}{BeO + MgO} \leq 1$. The introduction of 15% MgO (at a total BeO + MgO content $\leq 27.5\%$ and an MgO content greater than the BeO)

TABLE 1

Oxide content	Composition of glasses, %						Li₂O+MeO / B₂O₃	O/B	Glass formation and crystallization behavior
	weight			molecular					
	B_2O_3	Li_2O	MeO	B_2O_3	Li_2O	MeO	$\dfrac{Li_2O+MeO}{B_2O_3}$	$\dfrac{O}{B}$	
From	70,0	7,5	$\dfrac{BeO}{2,4}$	48,0	14,0	$\dfrac{BeO}{5,5}$	1,08	2,05	Glasses do not crystallize on cooling if the ratio $\dfrac{Li_2O}{BeO} > 1$
To	90,0	21,0	12,6	78,7	40,0	24,0	0,27	1,63	
From	80,0	2,5	$\dfrac{MgO}{1,0}$	66,3	5,3	$\dfrac{MgO}{1,5}$	0,52	1,76	Glasses do not crystallize on cooling at $\dfrac{Li_2O}{MgO} < 1$
To	91,5	10,0	10,0	83,0	21,0	16,0	0,21	1,6	
From	57,5	2,5	$\dfrac{BeO+MgO}{2,5+1,0}$	37,5	4,5	$\dfrac{BeO+MgO}{5,5+1,3}$	1,66	2,34	At $\dfrac{LiO}{BeO+MgO} > 1$ all the glasses were clear
To	91,5	15,0	15+15,0	82,0	28,0	28+21,0	0,22	1,61	At $\dfrac{Li_2O}{BeO+MgO} < 1$ the glasses were clear at MgO > BeO
From	66,0	2,3	$\dfrac{CaO}{5,0}$	60,0	5,0	$\dfrac{CaO}{5,0}$	0,7	1,91	Glasses did not crystallize at $\dfrac{Li_2O}{MeO} < 1$
To	90,0	20,0	31,0	84,0	36,5	35,0	0,2	1,6	
From	58,0	2,2	$\dfrac{ZnO}{5,0}$	60,0	5,0	$\dfrac{ZnO}{4,0}$	0,74	1,87	The same
To	90,0	20,0	40,0	85,0	37,0	35,0	0,18	1,59	

TABLE 1 (continued)

| Oxide content | Composition of glasses, % weight | | | molecular | | | $\dfrac{Li_2O+MeO}{B_2O_3}$ | $\dfrac{O}{B}$ | Glass formation and crystallization behavior |
	B_2O_3	Li_2O	MeO	B_2O_3	Li_2O	MeO			
From	52,0	1,8	$\dfrac{SrO}{5,0}$	60,0	5,0	$\dfrac{SrO}{3,0}$	1,67	1,83	Glasses did not crystallize at $\dfrac{Li_2O}{MeO} < 1$
To	90,0	20,0	46,0	86,0	37,3	35,0	0,17	1,58	
From	47,0	1,7	$\dfrac{CdO}{5,0}$	60,0	5,0	$\dfrac{CdO}{2,5}$	0,67	1,83	The same
To	90,0	20,0	51,0	86,0	37,5	35,0	0,16	1,58	
From	43,0	1,6	$\dfrac{BaO}{5,0}$	60,0	5,0	$\dfrac{BaO}{1,8}$	0,67	1,83	The same
To	90,0	20,0	55,2	87,0	37,5	35,0	0,16	1,58	
From	34,0	1,3	$\dfrac{PbO}{5,0}$	60,0	5,0	$\dfrac{PbO}{1,3}$	0,67	1,83	The same
To	90,0	20,0	64,3	87,5	37,5	35,0	0,16	1,56	
From	75,0	$\dfrac{BeO}{1,0}$	$\dfrac{MgO}{0}$	—	—	—	—	—	Clear glasses were not obtained
To	99,0	15,0	10,0	—	—	—	—	—	

10

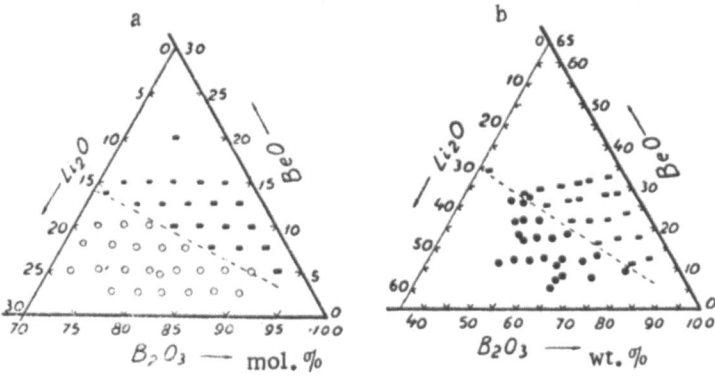

Fig. 1. $B_2O_3 - Li_2O - BeO$ systems: a) compositions in weight %; b) compositions in mol. %. □) glasses tending to crystallize on cooling the melt, ●, ○) clear glasses; ----) limit of glass-formation region (clear glasses).

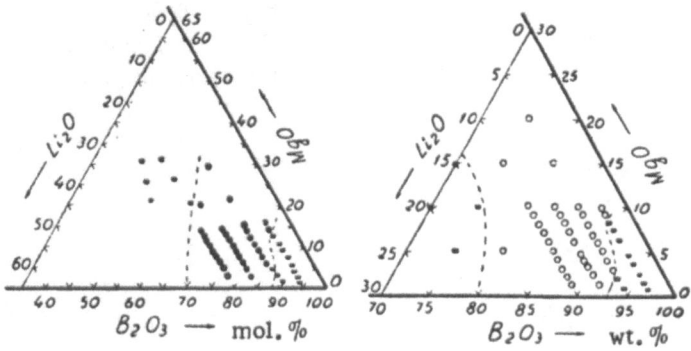

Fig. 2. $B_2O_3 - Li_2O - MgO$ system: □) glasses tending to crystallize on cooling the melt, ●, ○) clear glasses, ----) limits of glass-formation region (clear glasses).

made it possible to obtain glasses which did not crystallize on cooling, even when the glass contained only 2% of Li_2O.

Thus, the introduction of MgO considerably extended the glass-formation region and reduced the tendency of the glasses to crystallize on cooling. The behavior of MgO in glasses was found to be similar to Li_2O to a certain extent. However, glasses containing more than 25% of BeO + MgO showed a tendency to crystallize if the BeO content was equal to or greater than that of MgO.

This again confirms the difference in the behavior of BeO and MgO in the formation of the given borate glasses.

The primary phase in the crystallization of borate glasses on cooling, especially in high-boron and low-alkali ones, was B_2O_3, which on the surface rapidly changed into boric acid and also beryllium and magnesium borates, depending on the predominance of one or the other oxide in the glass.

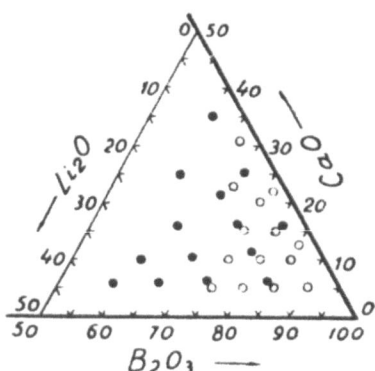

Fig. 3. $B_2O_3-Li_2O-CaO$ system:
$\bigcirc$) wt. %, $\bullet$) mol. %.

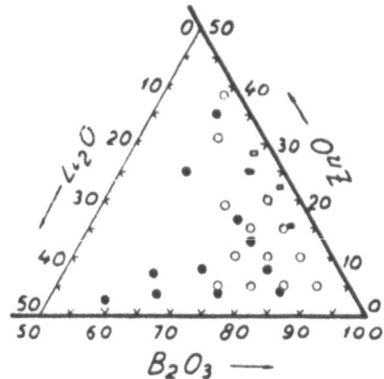

Fig. 4. $B_2O_3-Li_2O-ZnO$ system:
$\bigcirc$) wt. % , $\bullet$) mol. % , $\blacksquare$) crystallization of glass.

a

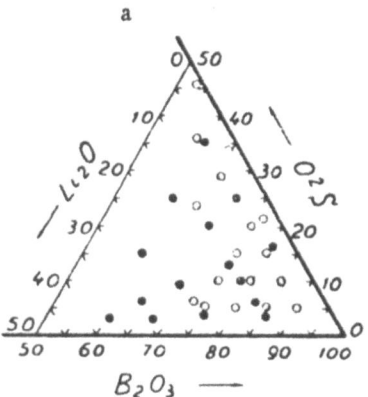

Fig. 5. $B_2O_3-Li_2O-SrO$ system:
$\bigcirc$) wt. % , $\bullet$) mol. %.

b

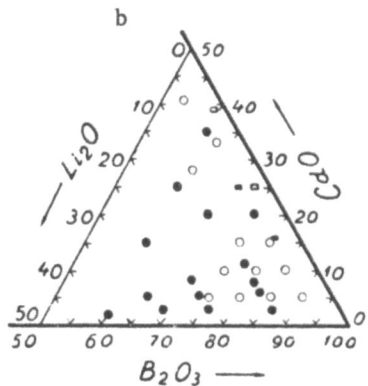

Fig. 6. $B_2O_3-Li_2O-CdO$ system:
$\bigcirc$) wt. %, $\bullet$) mol. %, $\blacksquare$) crystallization of glass.

Liquation on cooling was frequently observed in the same high-boron and low-alkali glasses when BeO and MgO were present in them together.

In the other systems, only glasses with a high B_2O_3 content (above 70 mol. %) and glasses with a low alkali content ($Li_2O \sim 2\%$) crystallized on cooling, i.e., those in which it can be assumed that a definite amount of unbound B_2O_3 was present.

The temperature and technological conditions of melting and working the glasses depend both on the B_2O_3 and Li_2O content of the glass and on the nature and content of the oxides MeO. Melts of high-boron glasses have a very low viscosity and a short range of working temperature, i.e., the glasses are "short." Beryllium and magnesium glasses are comparatively "short"; they mold and roll well but draw with difficulty and are blown with even greater difficulty. The other glasses had a relatively satisfactory working range and were worked normally.

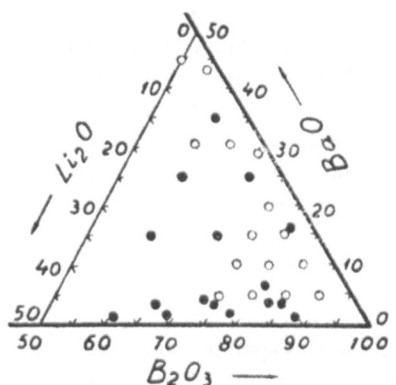

Fig. 7. $B_2O_3-Li_2O-BaO$ system. O) wt.%, ●) mol.%, ■) crystallization of glass.

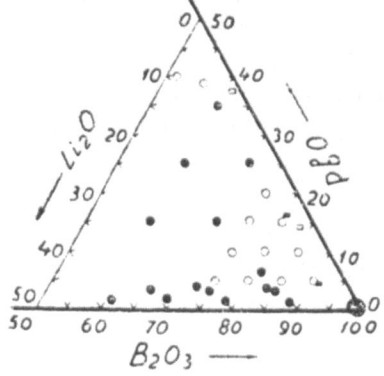

Fig. 8. $B_2O_3 - Li_2O - PbO$ system.

The greatest tendency toward volatilization on melting was shown by B_2O_3 at 1 to 5 wt.% (average 2-3%) and this should be considered in making up the charge. As stated above, the degree of volatilization depended to a considerable extent on the conditions of melting the charge. The molecular-percent compositions of clear glasses are presented in the appendix in Tables 9 and 13-20. The melting technology was also checked under factory conditions.

Since a procedure for chemical analysis of ternary and more complex glasses of the type $B_2O_3 - Li_2O - MeO$ had not been developed, in 1939 together with N. M. Neshchadimova and A. A. Sokolova we developed a method for chemical analysis of glasses with the composition $B_2O_3 - Li_2O - BeO$ and in 1953, together with T. I. Leizerov, methods of analyzing glasses with the composition $B_2O_3 - Li_2O - MgO$ and $B_2O_3 - Li_2O - BeO - MgO$, and also the same glasses with SiO_2 and Al_2O_3 added.

Glasses with the Composition $B_2O_3 - Li_2O - BeO$

The oxides were determined from two separate samples: the BeO and Li_2O were determined in one and the B_2O_3 in the other.

For the determination of BeO and Li_2O, a sample of glass (0.5-0.7 g) was placed in a platinum dish and moistened slightly with water. Then 2-3 ml of H_2SO_4 (density 1.84) was added and 15 ml of HF carefully introduced. The contents of the dish were evaporated in a current of air on an electric hotplate until SO_3 vapor appeared (100-110°). The dish was then taken from the hotplate and allowed to cool. The walls of the dish were washed with a small amount of water from a wash bottle, 5 ml of HF added, and the mixture again evaporated until SO_3 vapor appeared. This operation was again repeated (for complete removal of the B_2O_3). On appearance of thicker SO_3 vapor, the dish was transferred to a hotplate for stronger heating (up to 300°) and the mixture evaporated to dryness. The contents of the dish were then sintered slightly over a small burner flame (until the vapor evolution ceased). The dish was cooled, to the dry residue in it were added 10-20 ml of water and 10-15 ml of HCl (density 1.19) and the dish was heated on an electric hotplate (100°) until the residue dissolved completely. The clear solution obtained was transferred to a 250- to 300-ml beaker. The solution was boiled until the volume had been reduced to approximately 150 ml; carbon-dioxide-free ammonia was added to it dropwise to a yellow color with Methyl red. The precipitate of beryllium hydroxide thus produced

13

was allowed to settle and then collected on an ashless filter. The precipitate was washed three to four times with a hot, 1% solution of NH_4NO_3, then transferred together with the filter to the beaker in which it had been precipitated and dissolved in hot HCl (1:1). The solution was diluted to 150 ml with water, heated to boiling together with the pulp from the filter, and the beryllium hydroxide precipitated for a second time with ammonia to Methyl red. The precipitate was filtered and washed as in the first case until the reaction for Cl^- disappeared. The moist precipitate and the filter were placed in a weighed platinum crucible, the filter carefully ashed and the precipitate fired in a muffle furnace at 1000-1050°. The crucible was then cooled for 15 minutes in a desiccator over sulfuric acid (BeO is hygroscopic) and rapidly weighed. The crucible was tightly closed all the time with a platinum lid. The firing was repeated until the precipitate had a constant weight. As a result, we obtained the weight of BeO in %.

$$\frac{\text{weight of precipitate} \cdot 100}{\text{weight of glass sample}} = \% \text{ BeO}$$

For the determination of Li_2O, the filtrate from the BeO was evaporated to a volume of 50-80 ml, checking the completeness of beryllium hydroxide precipitation. If the precipitation was incomplete, a white, flocculent precipitate appeared when the filtrate was evaporated. In this case, the precipitate was collected on a small ashless filter, washed as described above, and added to the main precipitate in the crucible. The filtrate, concentrated in this way was transferred to a large silica dish and evaporated to dryness on a water bath. The evaporation process was very prolonged. Evaporation in a platinum dish is not recommended as the Cl^- and NO_3^- ions present have a harmful effect on the dish.

After evaporation of the solution, the dish was transferred to a drying cupboard (90-100°) and dried until the vapor of ammonium salts appeared and the temperature was then gradually raised to 120-125°. The dish was heated in a current of air with gentle flame from a gas burner until the ammonium salts had been removed completely. The cooled residue was dissolved in a few ml of water and the solution filtered into a weighed platinum dish.

To the filtrate in the dish was added 1-2 drops of concentrated sulfuric acid, the solution evaporated to dryness on a water bath, and the residue dried on an electric hotplate and then fired on a burner, at first carefully, then more strongly, until the residue melted. Firing was repeated until the precipitate reached constant weight. The dish was covered with a glass to prevent losses from cracking on cooling, placed in a desiccator and weighed when cool. As a result, we obtained the weight of Li_2O in %:

$$\frac{\text{weight of } Li_2SO_4 \text{ precipitate} \cdot 0.2718 \cdot 100}{\text{weight of glass sample}} = \% \text{ } Li_2O$$

For the determination of B_2O_3, a sample of 0.3-0.9 g of finely powdered glass was placed in a 500-ml flat-bottomed flask, 25-30 ml of HCl (1:1) added, and the mixture left for several hours in the cold until the glass dissolved. To the solution was added barium chloride solution (3 g of $BaCl_2$ to 10 ml of water) and 1-2 drops of Methyl orange and the mixture then neutralized to a yellow color with sodium carbonate. The flask was closed with a stopper with a reflux condenser. The flask with the reflux condenser was placed on an electric hotplate and the solution heated to boiling and boiled for 5 minutes. The flask was then cooled and the solution transferred to a 500-ml standard flask, diluted to the mark with water, and left for several hours. The solution was then filtered into two 200-ml standard flasks and parallel determinations carried out on the two samples.

14

The whole of the solution from a standard (200-ml) flask was transferred again to a 500-ml flat-bottomed flask, a few drops of dilute hydrochloric acid added (to an acid reaction to Methyl orange) and the solution boiled under reflux for 5 minutes to remove CO_2. After cooling, the solution was exactly neutralized to Methyl orange with 0.1 N NaOH. Then 5-10 drops of phenolphthalein and 10 ml of invert sugar were added and the solution titrated with 0.1 N NaOH to a pink color, 10 ml portions of invert sugar added several times until the pink color did not change on addition of invert sugar, and the total amount of 0.1 N NaOH consumed in titration to phenolphthalein noted. The calculation was carried out according to the following formula:

$$\frac{\text{number of ml of 0.1 N NaOH} \cdot \text{titer for } B_2O_3 \cdot 100 \cdot 500}{\text{weight of glass sample} \cdot 200} = \% \, B_2O_3$$

Glasses with the Composition $B_2O_3 - Li_2O - BeO - MgO$

These were analyzed in the following way.

The B_2O_3 was removed from a sample of glass powder with the composition $B_2O_3 - Li_2O - BeO - MgO + Al_2O_3$ by treatment with methyl alcohol and concentrated H_2SO_4. The mixture was evaporated on a water bath not less than 5 times for complete removal of the B_2O_3. After removal of the B_2O_3, the residue contained excess H_2SO_4, which was removed by firing the residue on a sand bath. After removal of the excess H_2SO_4, the residue was dissolved in hydrochloric acid and then either the beryllium hydroxide or the beryllium hydroxide and aluminum hydroxide precipitated with ammonia. The beryllium and aluminum hydroxide precipitate was collected and the MgO in the filtrate determined by precipitation with ammonium phosphate. For the complete separation of MgO, its determination was carried out by three precipitations.

The Li_2O in the filtrate from the MgO determination was determined by evaporation on a water bath, then conversion of the residue into Li_2SO_4 with dilute H_2SO_4 and subsequent firing in a muffle furnace at 1000-1100° to constant weight, which required quite a long time.

The B_2O_3 was removed from glass sample with the composition $B_2O_3 - Li_2O - BeO - MgO + Al_2O_3 + SiO_2$ by the method described above. After removal of the excess H_2SO_4, the dry residue was fused with $KNaCO_3$. The SiO_2 was precipitated and collected by the usual method. Beryllium and aluminum hydroxides were precipitated from the filtrate with ammonia and after their removal, the MgO was precipitated with ammonium phosphate. The Li_2O was determined on a separate sample by the method of Berzelius [138].

The beryllium and aluminum hydroxides were separated by the following methods:

1. The $Be(OH)_2$ and $Al(OH)_3$ precipitates were dissolved in excess alkali with the formation of beryllates and aluminates. Heating the alkaline solution reprecipitated the $Be(OH)_2$ in contrast to the $Al(OH)_3$.

2. The $Be(OH)_2$ and $Al(OH)_3$ precipitates were dissolved in ammonium carbonate. The white precipitate of $Be(OH)_2$ dissolved readily in excess solvent in contrast to $Al(OH)_3$.

THERMOCHEMICAL PROCESSES AND REACTIONS OF GLASS FORMATION

The processes and reactions of glass formation have been studied only for certain binary boroalkali systems [52-59]; those of ternary and more complex boron systems remain unstudied. Literature data on the properties of H_3BO_3 and B_2O_3 [132-141] are very contradictory. For example, according to Berg [132] H_3BO_3 dissociates to HBO_3 at 150° and to B_2O_3 at 200°. According to "The Chemist's Handbook" [139], H_3BO_3 decomposes at 185° and boils at 300°, having lost H_2O. According to Nikolaev [45], a loss in weight by H_3BO_3 is noticeable even at 100°. According to Taylor and Nekrasov [39, 41], B_2O_3 melts at 294° and according to "The Chemist's Handbook" [42] at 577°. Data on the dissociation temperatures of a number of components of the type $MeCO_3$ ($MgCO_3$, $CdCO_3$, etc.) are contradictory and frequently absent [139-141]. The problem of the structure of borate glasses is even less clear.

The purpose of our investigation was to establish the processes and reactions of glass formation in binary, ternary, and quaternary systems of the type $B_2O_3 - Li_2O$, $B_2O_3 - MeO$, $MeO - Li_2O$, and $B_2O_3 - Li_2O - MeO$ and also to determine more accurately the dissociation temperatures of the separate components.

The determinations were carried out by the differential thermal-analysis method on an apparatus of the Kurnakov type [141-152]. For the same compositions as in the thermal analysis, we carried out a parallel dynamic weighing of samples during heating and determined the weight loss. We obtained and analyzed thermograms and also weight curves during heating of the following components and mixtures of them in different proportions (molecular):

H_3BO_3, Li_2CO_3, $BeCO_3$, $MgCO_3$, $CaCO_3$, $SrCO_3$, $CdCO_3$, $BaCO_3$;

$H_3BO_3 + Li_2CO_3$ ($BeCO_3$, $MgCO_3$, $CaCO_3$, ZnO, $SrCO_3$ $CdCO_3$, $BaCO_3$, PbO);

$Li_2CO_3 + MeCO_3$ ($BeCO_3$, $MgCO_3$, $CaCO_3$ ZnO, $SrCO_3$, $CdCO_3$, $BaCO_3$, PbO);

$H_3BO_3 + Li_2CO_3 + MeCO_3$ ($BeCO_3$, $MgCO_3$ $CaCO_3$, ZnO, $SrCO_3$, $CdCO_3$. $BaCO_3$, PbO);

$H_3BO_3 + Li_2CO_3 + BeCO_3 + MgCO_3$;

$BeCO_3 + MgCO_3$.

We also obtained 16 electron pictures of 8 glasses with the compositions $B_2O_3 - Li_2O - BeO$, $B_2O_3 - Li_2O - MgO$ and $B_2O_3 - Li_2O - BeO - MgO$.

A summary of the processes and reactions of glass formation based on an analysis of the results of investigating the systems is presented in Table 2. The thermograms and weight loss curves are presented in Figs. 9-30.

TABLE 2

A Short Summary of the Temperature Ranges of the Main Thermochemical Processes and Reactions of Glass Formation

Components and mixtures of them	Main endothermal processes and effects	Exothermal processes	Temperature of total loss of volatile components, deg
H_3BO_3 (Fig. 9, 14)	75-100° – beginning of dissociation, 150° – endothermal effect – dissociation maximum, 300° – almost complete loss of volatile components, 515-525° – endothermal inflection – fusion of B_2O_3	—	350
$BeCO_3$ (Fig. 9, 14)	150° – beginning of decomposition, 325-350° – endothermal effect – decomposition maximum, 400-450° – almost complete loss of volatile components (up to 97%)	—	600
$MgCO_3$ (Fig. 9, 14)	300° – beginning of decomposition, 400° – endothermal effect – decomposition maximum, 500-550° – almost complete loss of volatile components (up to 95%)	—	700
Li_2CO_3 (Fig. 9, 14)	650° – beginning of decomposition, 725° – endothermal effect – decomposition maximum, 900° – almost complete loss of volatile components (> 90%)	—	950

TABLE 2 (continued)

Components and mixtures of them	Main endothermal processes and effects	Exothermal processes	Temperature of total loss of volatile components, deg
$CaCO_3$ (Fig. 23, 26)	550° – beginning of decomposition, 940° – end of decomposition	–	940
$SrCO_3$ (Fig. 25, 26)	850° – beginning of decomposition, 1100° – incomplete decomposition	–	>1100
$CdCO_3$ (Fig. 27, 30)	360° – beginning of decomposition, 700° – complete decomposition	–	700
$BaCO_3$ (Fig. 28, 30)	825° – beginning of decomposition, 1200° – incomplete decomposition	–	>1200
$4H_3BO_3 + BeCO_3$ (Fig. 10–12)	250° – endothermal effect – maximum of loss of volatile components, 525° – fusion of B_2O_3 and beginning of partial solution of mixture in melt and also reaction of B_2O_3 with BeO	650-675° – very small exothermal effect – formation of compound of the type $ABeO \cdot B_2O_3$	450
$3H_3BO_3 + BeCO_3$ (Fig. 10–12)	The same	The same, but a more considerable exothermal effect	450
$2H_3BO_3 + BeCO_3$ (Fig. 10–12)	The same	The same	470
$1.5H_3BO_3 + BeCO_3$ (Fig. 10–12)	The same	The same	500
$H_3BO_3 + BeCO_3$ (Fig. 10–12)	300-350° – endothermal effect – maximum of loss of volatile components	–	450

TABLE 2 (continued)

Components and mixtures of them	Main endothermal processes and effects	Exothermal processes	Temperature of total loss of volatile components, deg
$H_3BO_3 + 2BeCO_3$ (Fig. 10-12)	300-375° – the same	650 and 750-850° – exothermal effect – formation of compounds of the type $ABeO \cdot B_2O_3$	450
$H_3BO_3 + 3BeCO_3$ (Fig. 10-12)	300-375° – endothermal effect – maximum of loss of volatile components	650 and 750-850° – exothermal effect – formation of compounds of the type $ABeO \cdot B_2O_3$	450
$H_3BO_3 + 4BeCO_3$ (Fig. 10-12)	The same	The same	450
$H_3BO_3 + Li_2CO_3$ (Fig. 13, 14)	200-250° – endothermal effect – maximum of H_3BO_3 dehydration, 600° – decomposition of $LiCO_3$, 625-650° – fusion of mixture and formation of eutectic with the composition $Li_2O \cdot AB_2O_3$	725° – very small exothermal effect – formation of compounds of the type $Li_2O \cdot AB_2O_3$ ($Li_2O \cdot 3B_2O_3 + Li_2O \cdot 4B_2O_3$), according to literature sources [52]	600
$H_3BO_3 + 2Li_2CO_3$ (Fig. 13, 14)	The same	–	700
$2H_3BO_3 + Li_2CO_3$ (Fig. 13, 14)	The same	–	600
$4H_3BO_3 + MgCO_3$ (Fig. 15-17)	250-300° – endothermal effect – dehydration of H_3BO_3 and beginning of $MgCO_3$ decomposition, 475° – very small endothermal effect – decomposition of $MgCO_3$, 660-675° – endothermal effects – formation of eutectics of MgO with B_2O_3	–	650

19

TABLE 2 (continued)

Components and mixtures of them	Main endothermal processes and effects	Exothermal processes	Temperature of total loss of volatile components, deg
$3H_3BO_3 + MgCO_3$ (Fig. 15-17)	250-300° – endothermal effect – dehydration of H_3BO_3 and beginning of $MgCO_3$ decomposition, 475° – very small endothermal effect – decomposition of $MgCO_3$, 660-675° – endothermal effects – formation of eutectics of MgO with B_2O_3	635 and 760° exothermal effects – formation of magnesium borates of the diborate and inderite	650
$2H_3BO_3 + MgCO_3$ (Fig. 15-17)	The same	760 and 825-850° – exothermal effect – formation of compounds of pinnoite and inderite type	650
$1.5H_3BO_3 + MgCO_3$ (Fig. 15-17)	The same	635, 690 and 825° – the same exothermal effects	650
$H_3BO_3 + MgCO_3$ (Fig. 15-17)	The same	635-760° – exothermal effects	650
$H_3BO_3 + 2MgCO_3$ (Fig. 15-17)	The same	The same	650
$H_3BO_3 + 3MgCO_3$ (Fig. 15-17)	The same	The same	650
$H_3BO_3 + 4MgCO_3$ (Fig. 15-17)	The same	The same	650

20

TABLE 2 (continued)

Components and mixtures of them	Main endothermal processes and effects	Exothermal processes	Temperature of total loss of volatile components, deg
H_3BO_3 + $CaCO_3$ (Fig. 23, 26)	$100-350°$ — decomposition of H_3BO_3, $375-975°$ — production of liquation $CaO - B_2O_3$, $785°$ — $CaO - B_2O_3$ eutectic, $900-960°$ — formation of compounds of the type $CaO \cdot B_2O_3$ and $CaO \cdot 2B_2O_3$	—	800
H_3BO_3 + ZnO (Fig. 24, 26)	$100-525°$ — decomposition of H_3BO_3 and fusion of B_2O_3, $800-900°$ — formation of compounds $ZnO \cdot AB_2O_3$	—	300
H_3BO_3 + $SrCO_3$ (Fig. 25, 26)	$100-525°$ — see above, $650°$ — formation of an $SrO \cdot AB_2O_3$ melt, $790-880°$ — formation of compounds of the type $SrO \cdot AB_2O_3$	—	950
H_3BO_3 + $CdCO_3$ (Fig. 27-30)	$100-525°$ — see above, $720-750°$ — formation of compounds of the type $CdO \cdot B_2O_3$ and $2CdO \cdot 3B_2O_3$	—	700
H_3BO_3 + $BaCO_3$ (Fig. 28, 30)	$100-525°$ — see above, $800, 925$ and $990°$ — formation of the compounds $BaO \cdot B_2O_3$, $3BaO \cdot B_2O_3$ and $BaO \cdot 3B_2O_2$	—	950

TABLE 2 (continued)

Components and mixtures of them	Main endothermal processes and effects	Exothermal processes	Temperature of total loss of volatile components, deg
H_3BO_3 + PbO (Fig. 29, 30)	100-525° — see above, 560° — formation of a eutectic of PbO and B_2O_3, 650° — fusion of mixture and production of liquation	—	300
$BeCO_3$ + $MgCO_3$ (Fig. 19)	150-500° — almost complete decomposition of $BeCO_3$ and $MgCO_3$. Reactions between them not established	—	500
Li_2CO_3 + $BeCO_3$ (Fig. 19, 20)	125-415° — decomposition of $BeCO_3$, 540° — weak endothermal effect, 670-770° — decomposition of Li_2CO_3, 790° — very small endothermal effect	870° — exothermal effect	950
Li_2CO_3 + $MgCO_3$ (Fig. 21)	360-475° — decomposition of $MgCO_3$, 540° — weak endothermal effect, 670-770° — decomposition of Li_2CO_3	900° — exothermal effect — formation of compounds of Li_2O with MgO	950
Li_2CO_3 + $CaCO_3$ (Fig. 23, 26)	900-980° — formation of melt of CaO with Li_2O and separation of compound $Li_2O \cdot 2CaO$	—	900
Li_2CO_3 + ZnO (Fig. 24, 26)	800° — beginning of formation of melt of ZnO with Li_2O	—	900
Li_2CO_3 + $SrCO_3$ (Fig. 25, 26)	975° — formation of melt of SrO with Li_2O	—	950

TABLE 2 (continued)

Components and mixtures of them	Main endothermal processes and effects	Exothermal processes	Temperature of total loss of volatile components, deg
$Li_2CO_3 + CdCO_3$ (Fig. 27, 30)	800° — formation of eutectic of Li_2O with CdO	—	850
$Li_2CO_3 + BaCO_3$ (Fig. 28, 30)	925° — eutectic of Li_2O and BaO	—	>1100
$Li_2CO_3 + PbO$ (Fig. 29, 30)	970° — formation of compound $Li_2O \cdot PbO$	—	800
$H_3BO_3 + Li_2CO_3 + BeCO_3$ (Fig. 18, 19)	150° — endothermal effect — maximum of H_3BO_3 dehydration, 300° — endothermal effect — decomposition of $BeCO_3$, 525° — fusion of B_2O_3 550° — endothermal effect — decomposition of Li_2CO_3, 650–750° — formation of eutectic — lithium and beryllium borates	800° — formation of beryllium and lithium borates — exothermal effect	750
$2H_3BO_3 + Li_2CO_3 + BeCO_3$ (Fig. 18, 19)	The same	650 and 800° — the same exothermal effects	650
$3H_3BO_3 + Li_2CO_3 + BeCO_3$ (Fig. 18, 19)	The same	The same	650

23

TABLE 2 (continued)

Components and mixtures of them	Main endothermal processes and effects	Exothermal processes	Temperature of total loss of volatile components, deg
H_3BO_3 + Li_2CO_3 + $MgCO_3$ (Fig. 18, 19)	150-200° — maximum of H_3BO_3 dehydration, 415-450° — maximum of $MgCO_3$ decomposition, 575° — endothermal effect — beginning of decomposition and formation of eutectic of MgO and B_2O_3, 660-715° — formation of eutectic of B_2O_3 with Li_2O and MgO and also magnesium and lithium borates	625° — exothermal effect — formation of magnesium and lithium borates	800
$2H_3BO_3$ + Li_2CO_3 + $MgCO_3$ (Fig. 18, 19)	The same	The same	800
$3H_3BO_3$ + Li_2CO_3 + $MgCO_3$ (Fig. 18, 19)	The same	The same	700
H_3BO_3 + Li_2CO_3 + $CaCO_3$ (Fig. 23, 26)	525° — eutectic of B_2O_3 with CaO 800-1000° — formation of compounds of B_2O_3 with CaO and Li_2O	—	900
H_3BO_3 + Li_2CO_3 + ZnO (Fig. 24, 26)	800-900° — formation of compound $ZnO \cdot B_2O_3$ and also of B_2O_3 with Li_2O and ZnO	—	850
H_3BO_3 + Li_2CO_3 + $SrCO_3$ (Fig. 25, 26)	700-850° — formation of eutectics of B_2O_3 with Li_2O and partially with SrO 950° — reaction of B_2O_3 with SrO	—	~1000

TABLE 2 (continued)

Components and mixtures of them	Main endothermal processes and effects	Exothermal processes	Temperature of total loss of volatile components, deg
$H_3BO_3 + Li_2CO_3 + CdCO_3$ (Fig. 27, 30)	750° – formation of compounds $aCdO \cdot cB_2O_3$ 850° – formation of compounds of B_2O_3 with CdO and Li_2O	–	750
$H_3BO_3 + Li_2CO_3 + BaCO_3$ (Fig. 28, 30)	825-925° – formation of eutectics and compounds $aBaO \cdot bB_2O_3$, $Li_2O \cdot 2B_2O_3$, and also ternary borates	–	900
$H_3BO_3 + Li_2CO_3 + PbO$ (Fig. 29, 30)	800° and above – formation of eutectics and binary and ternary compounds of B_2O_3 with PbO and Li_2O	–	700
$H_3BO_3 + Li_2CO_3 + BeCO_3 + MgCO_3$ (Fig. 20–22)	150-200° – maximum of H_3BO_3 dehydration, 315° – maximum of $BeCO_3$ decomposition, 400-450° – endothermal effect – decomposition of $MgCO_3$, 575° – decomposition of Li_2CO_3 and beginning of formation of eutectics of Li_2O with other oxides, 650-690° – formation of eutectics of oxides, see above, 715-790° – the same, see above	825° – exothermal effects – formation of borates and their separation from the melt	750
$2H_3BO_3 + Li_2CO_3 + BeCO_3 + MgCO_3$ (Fig. 20–22)	The same	The same	750

TABLE 2 (continued)

Components and mixtures of them	Main endothermal processes and effects	Exothermal processes	Temperature of total loss of volatile components, deg
$3H_3BO_3 + Li_2CO_3 + BeCO_3 + MgCO_3$ (Fig. 20–22)	The same	The same	700
$4H_3BO_3 + Li_2CO_3 + BeCO_3 + MgCO_3$ (Fig. 20–22)	The same	The same	700
$H_3BO_3 + 2Li_2CO_3 + BeCO_3 + MgCO_3$ (Fig. 20–22)	The same	The same	850
$2H_3BO_3 + 2Li_2CO_3 + BeCO_3 + MgCO_3$ (Fig. 20–22)	The same	The same	800
$3H_3BO_3 + 2Li_2CO_3 + BeCO_3 + MgCO_3$ (Fig. 20–22)	The same	The same	750
$4H_3BO_3 + 2Li_2CO_3 + BeCO_3 + MgCO_3$ (Fig. 20–22)	The same	The same	700

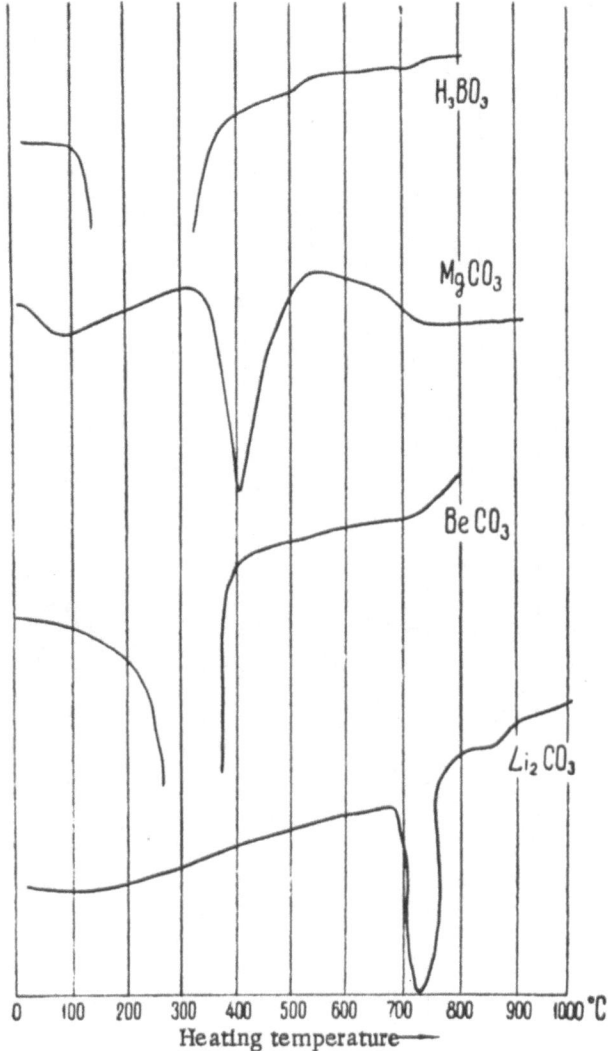

Fig. 9. Thermograms of H$_3$BO$_3$, BeCO$_3$, MgCO$_3$ and Li$_2$CO$_3$.

27

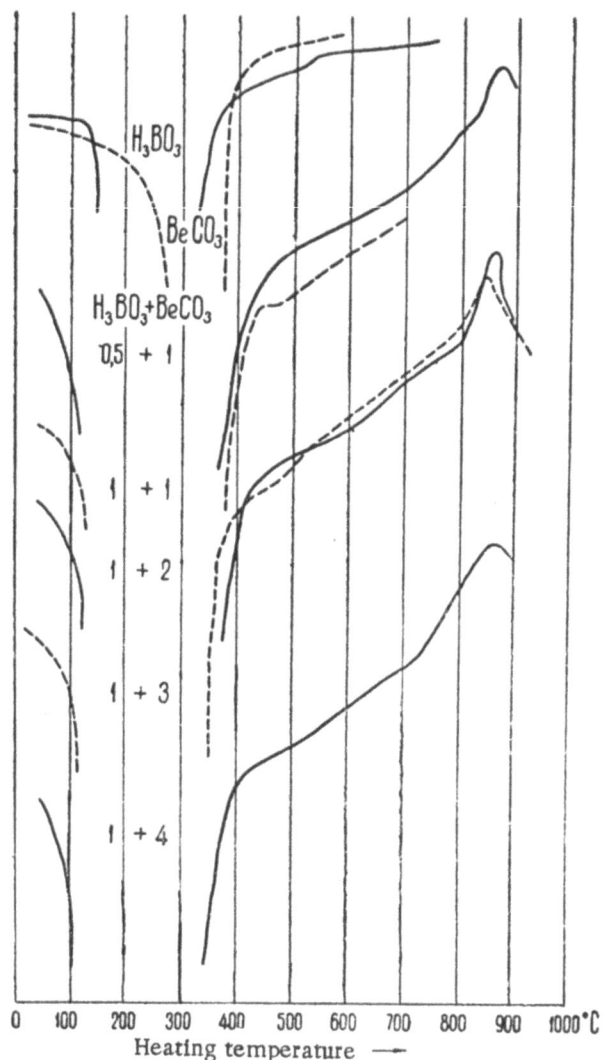

Fig. 10. Thermograms of H_3BO_3 + $BeCO_3$ mixtures.

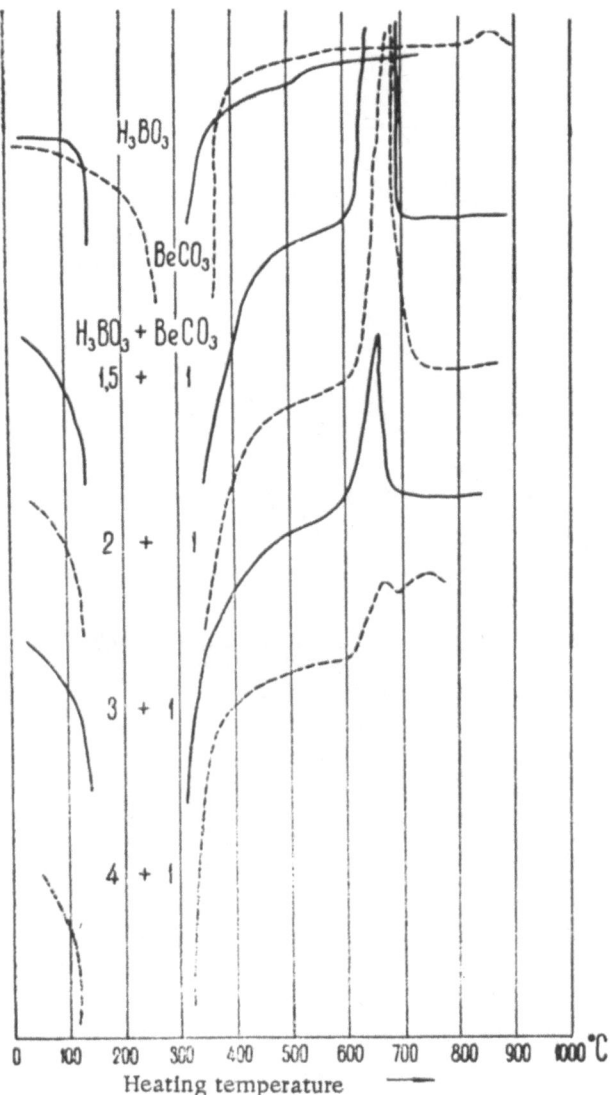

Fig. 11. Thermograms of H_3BO_3 + $BeCO_3$ mixtures (concluded).

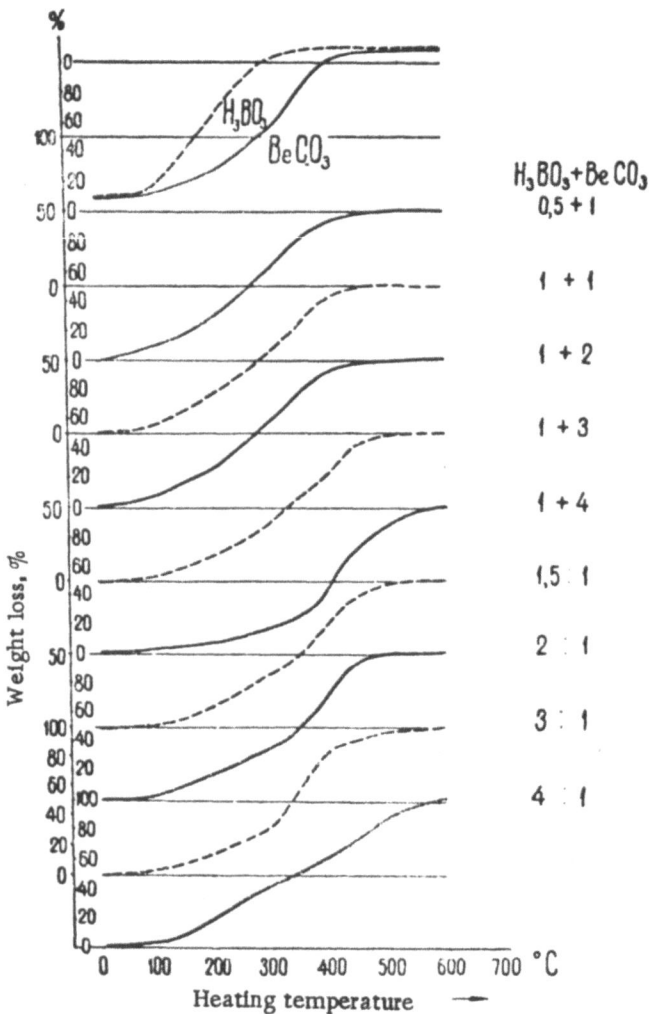

Fig. 12. Weight-loss curves on heating of H_3BO_3 + $BeCO_3$ mixtures.

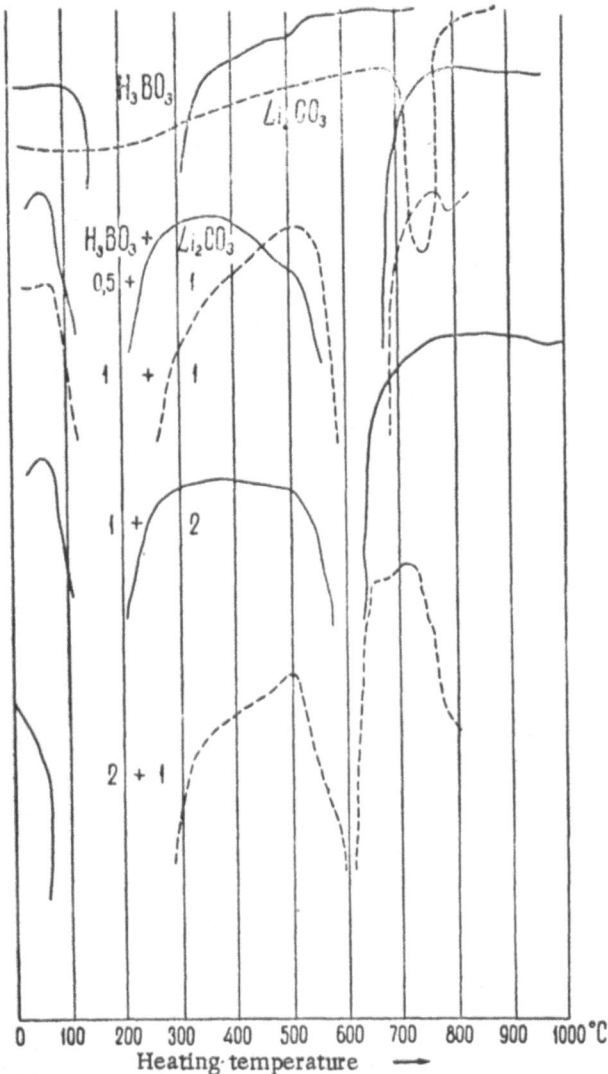

Fig. 13. Thermograms of H_3BO_3 + Li_2CO_3 mixtures.

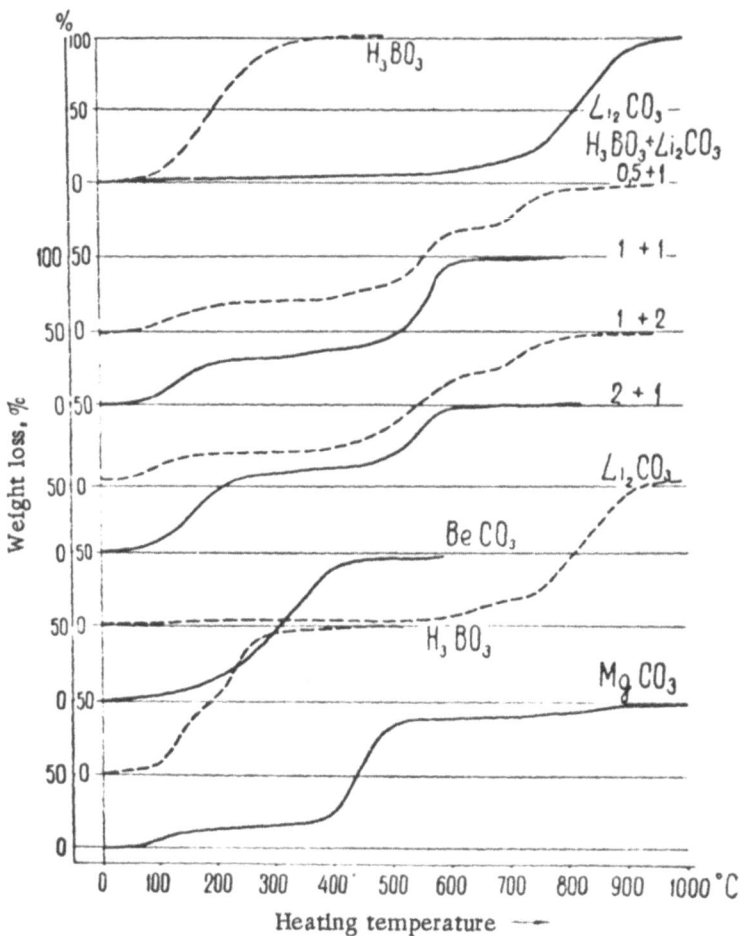

Fig. 14. Weight-loss curves on heating of H_3BO_3, $BeCO_3$, $MgCO_3$, Li_2CO_3 and $H_3BO_3 + Li_2CO_3$ mixtures.

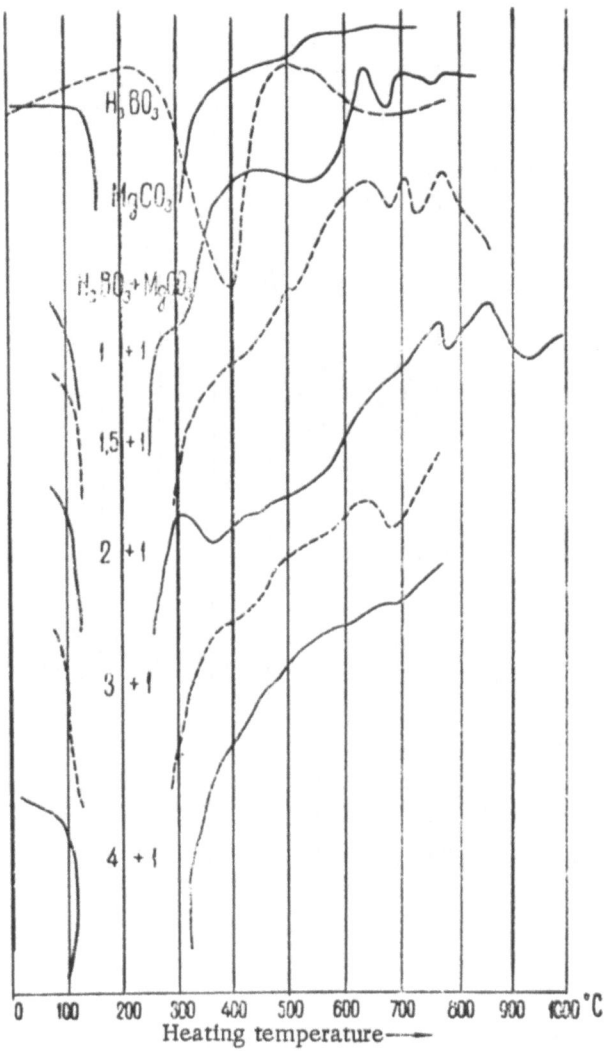

Fig. 15. Thermograms of H_3BO_3 + $MgCO_3$ mixtures.

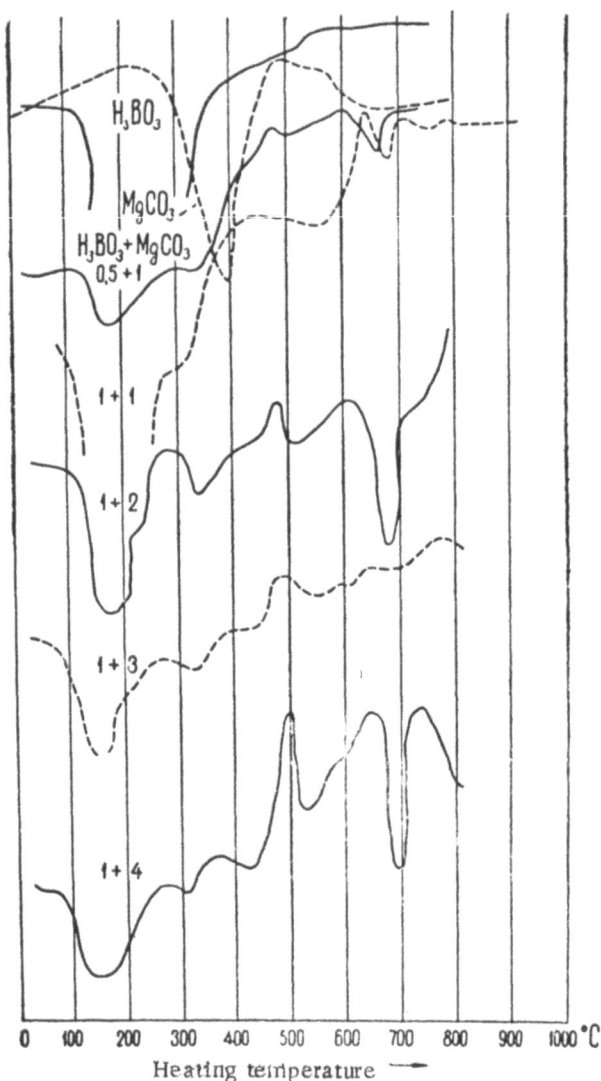

Fig. 16. Thermograms of H_3BO_3 + $MgCO_3$ mixtures (concluded).

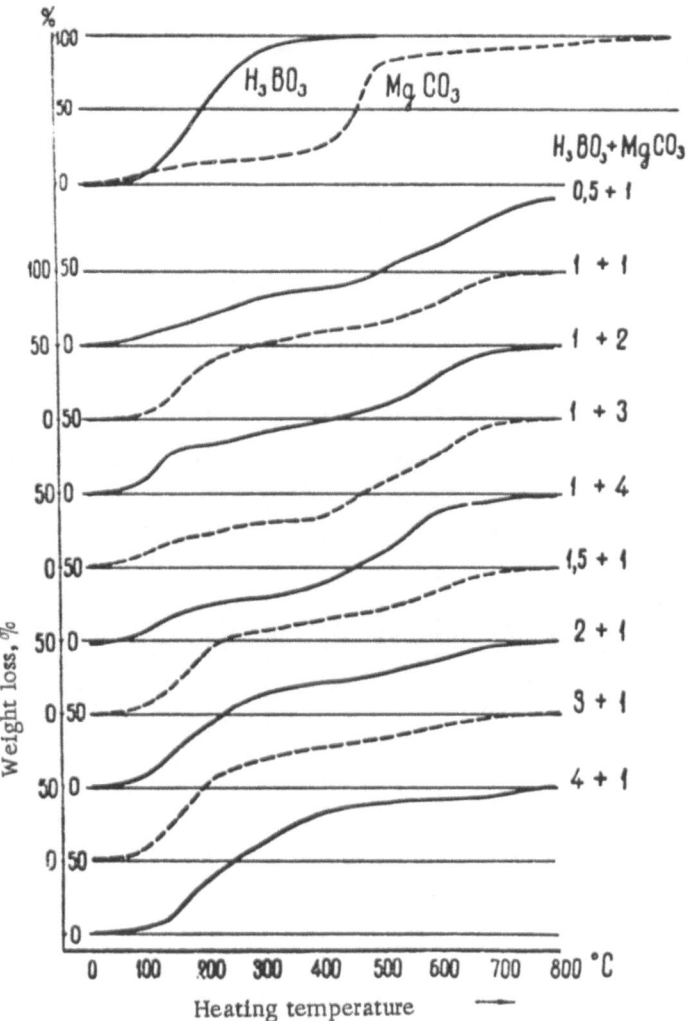

Fig. 17. Weight-loss curves on heating of H_3BO_3 + $MgCO_3$.

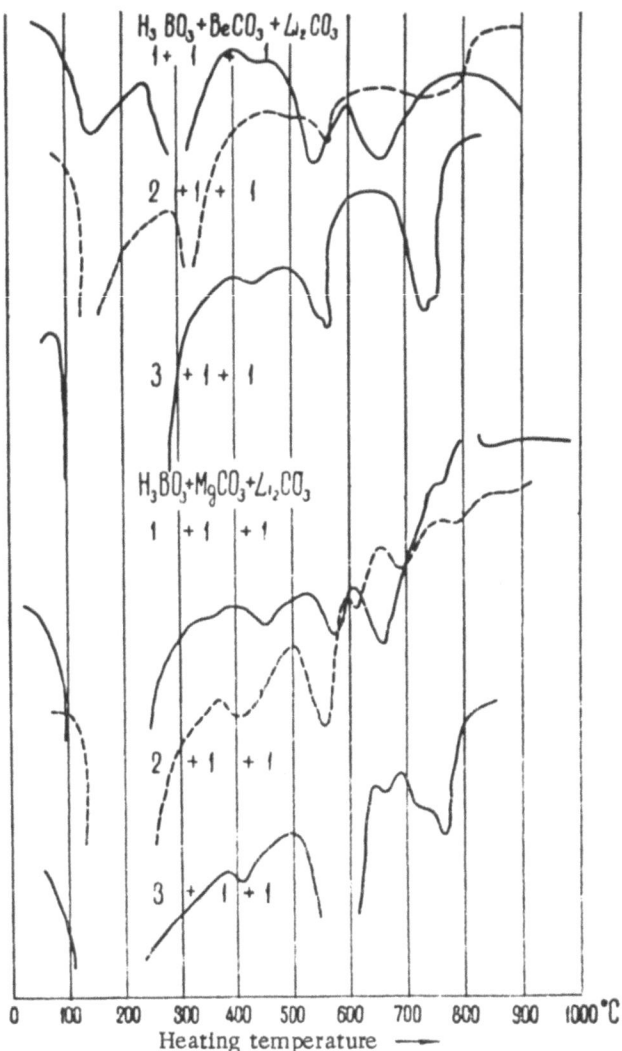

Fig. 18. Thermograms of H_3BO_2 + $BeCO_3$ ($MgCO_3$) + Li_2CO_3.

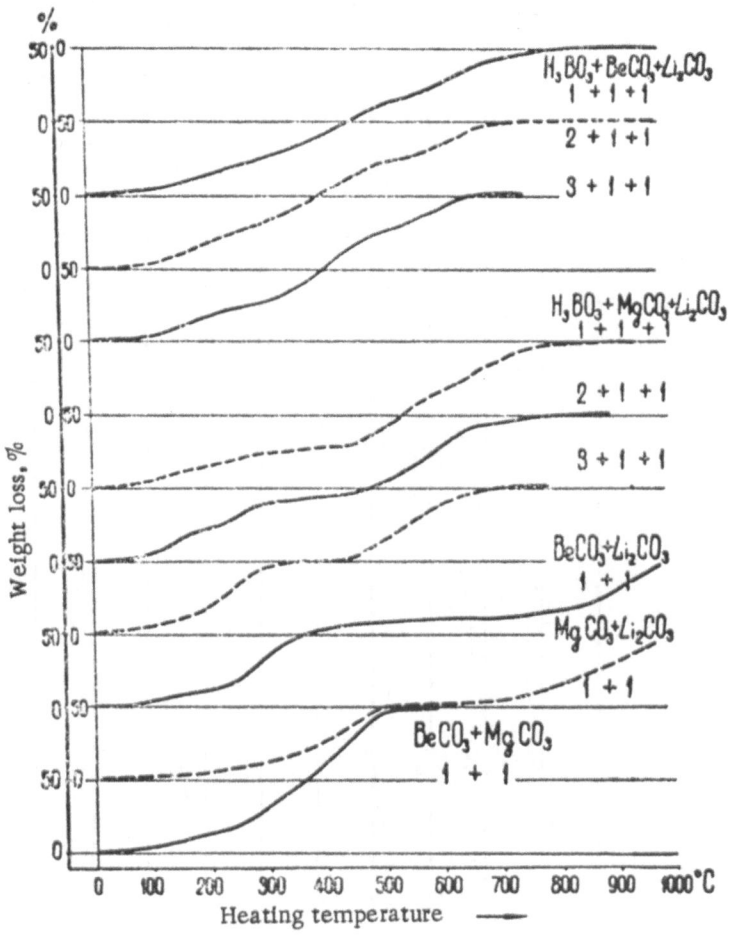

Fig. 19. Weight-loss curves on heating of H_3BO_3 + $BeCO_3$ ($MgCO_3$) +
+ Li_2CO_3, $BeCO_3$ ($MgCO_3$) + Li_2CO_3 and $BeCO_3$ + $MgCO_3$ mixtures.

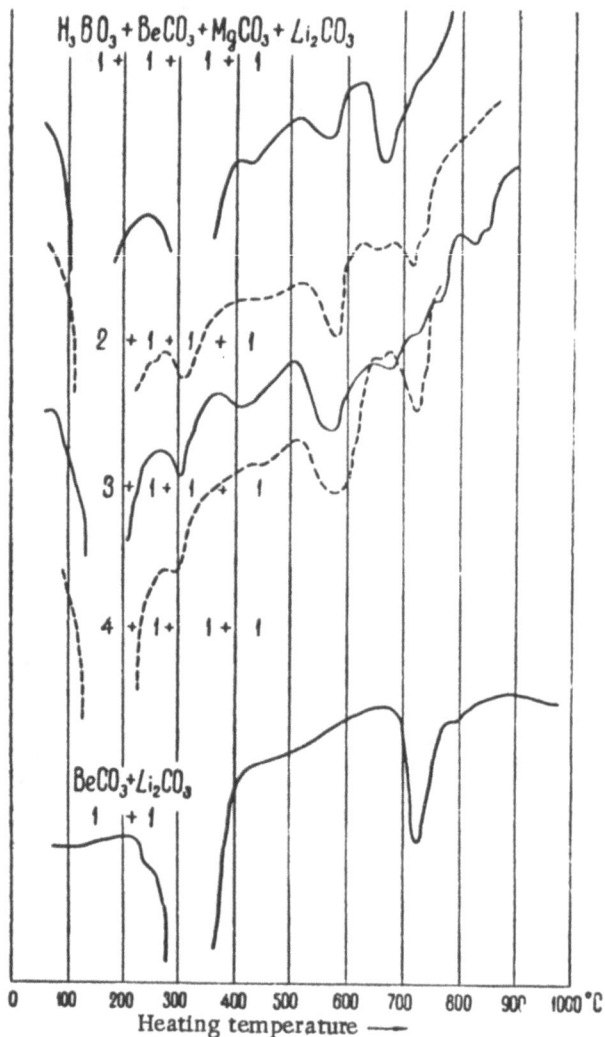

Fig. 20. Thermograms of H_3BO_3 + $BeCO_3$ + $MgCO_3$ + + Li_2CO_3 and also $BeCO_3$ + Li_2CO_3 mixtures.

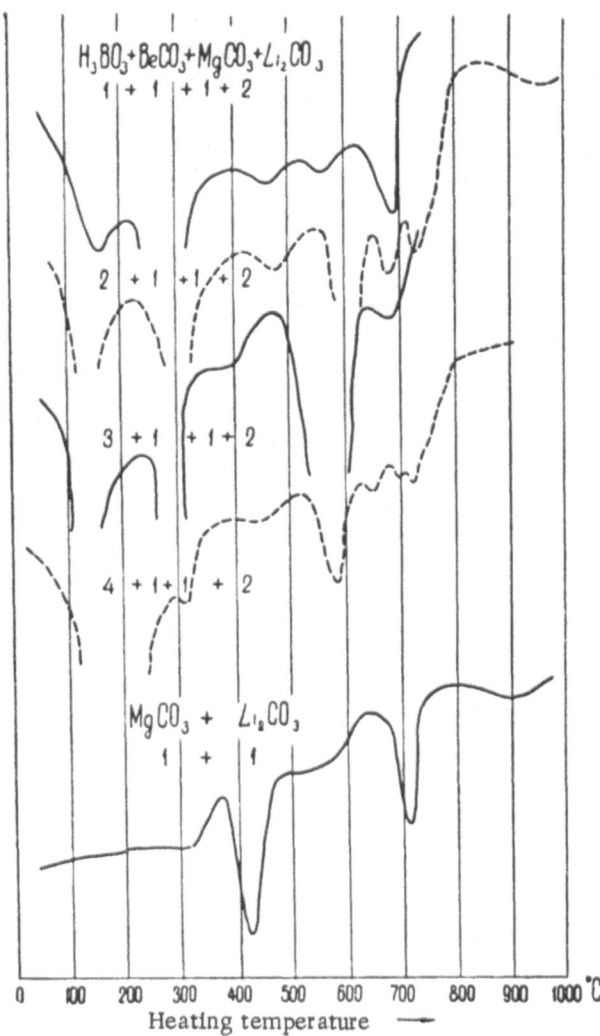

Fig. 21. Thermograms of H_3BO_3 + $BeCO_3$ + $MgCO_3$ + + Li_2CO_3 (concluded) and also $MgCO_3$ + Li_2CO_3 mixtures.

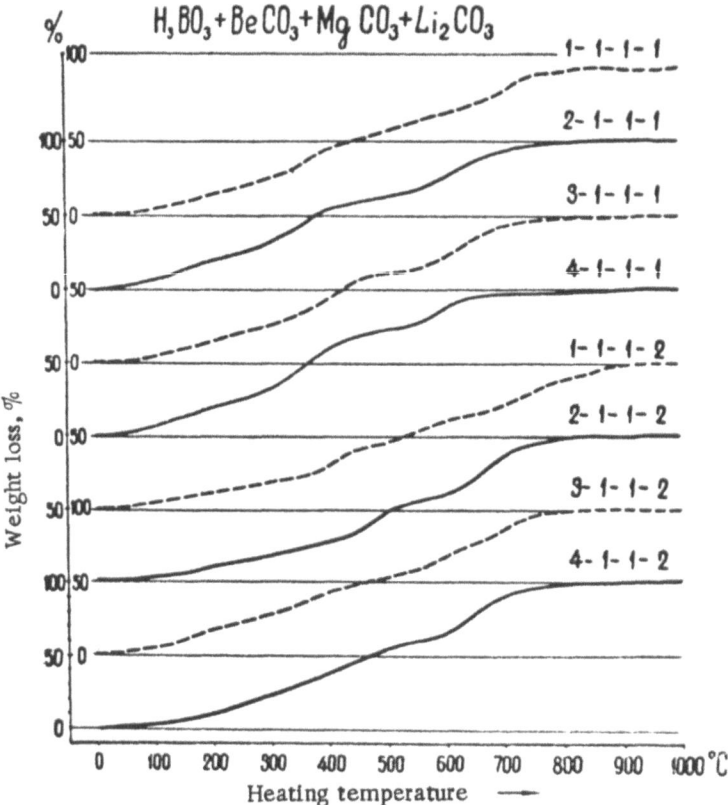

Fig. 22. Weight-loss curves on heating of H_3BO_3 + $BeCO_3$ + $MgCO_3$ + + Li_2CO_3 mixtures.

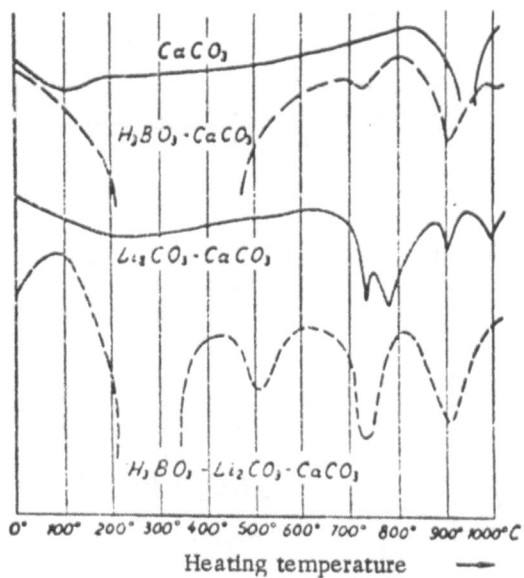

Fig. 23. Thermograms of CaCO₃ and CaCO₃ + H₃BO₃ (Li₂CO₃) and CaCO₃ + Li₂CO₃ + H₃BO₃ mixtures.

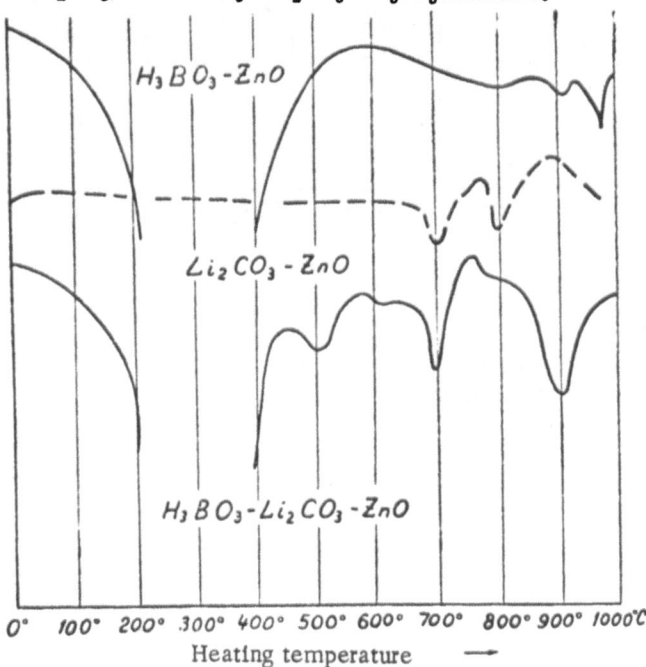

Fig. 24. Thermograms of ZnO + H₃BO₃ (Li₂CO₃) and H₃BO₃ + + Li₂CO₃ + ZnO mixtures.

41

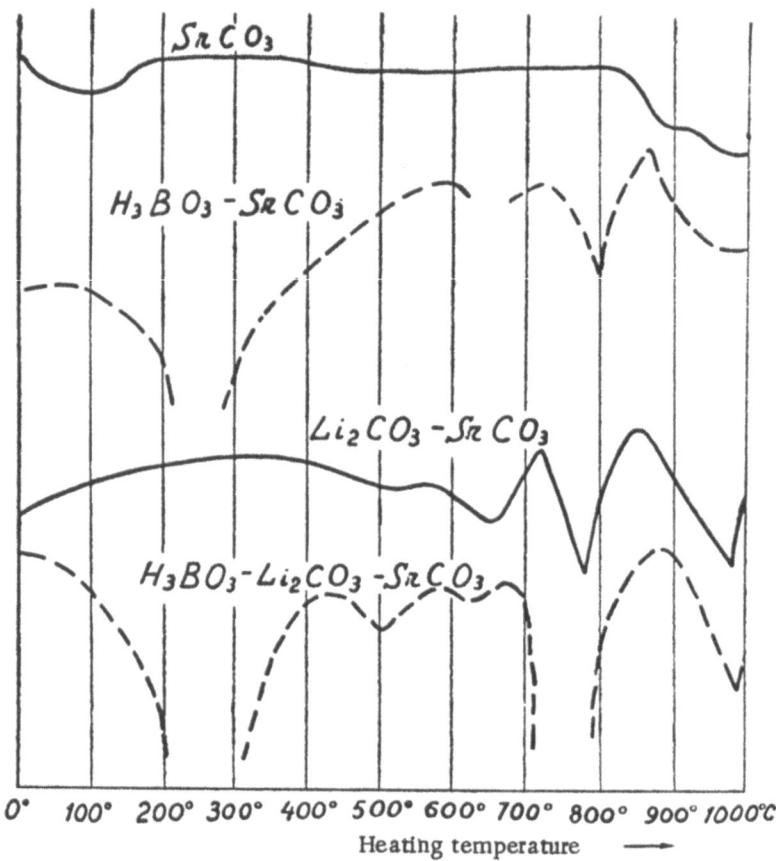

Fig. 25. Thermograms of SrCO₃ and SrCO₃ + H₃BO₃ (Li₂CO₃) and SrCO₃ + + Li₂CO₃ + H₃BO₃ mixtures.

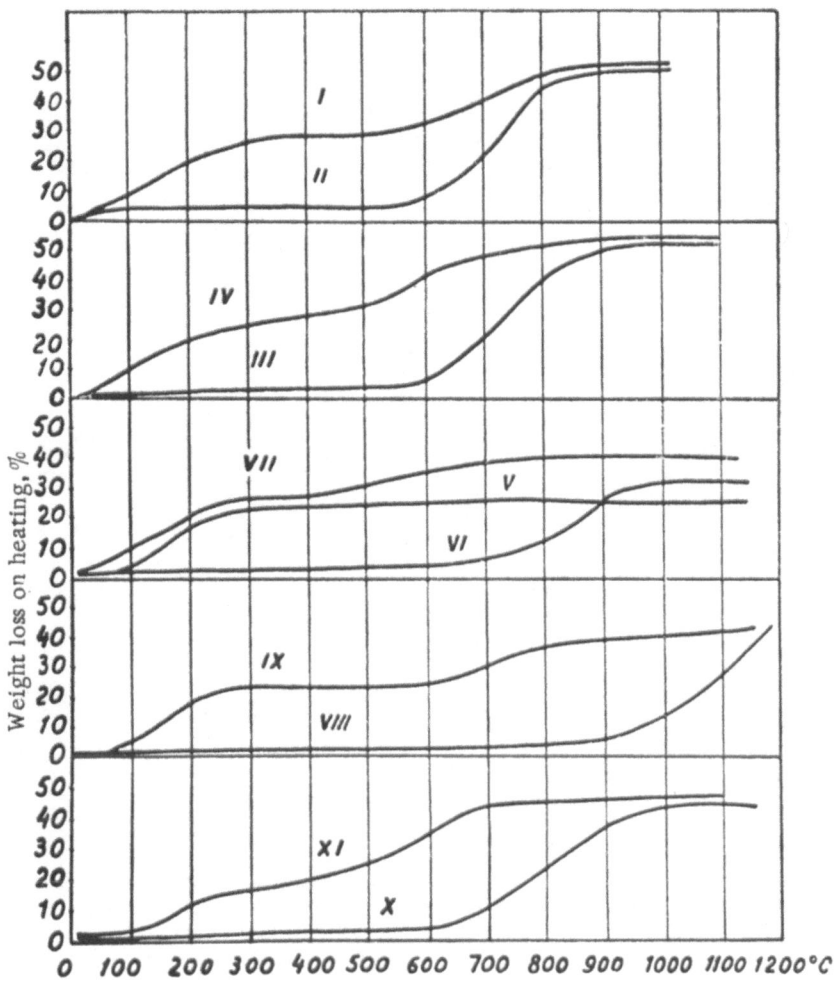

Fig. 26. Weight-loss curves on heating of the following components and their binary and ternary mixtures: $CaCO_3$ (ZnO, $SrCO_3$), Li_2CO_3 and H_3BO_3.

I. $H_3BO_3 + CaCO_3$

II. $CaCO_3$

III. $Li_2CO_3 + CaCO_3$

IV. $H_3BO_3 + Li_2CO_3 + CaCO_3$

V. $H_3BO_3 + ZnO$

VI. $Li_2CO_3 + ZnO$

VII. $H_3BO_3 + Li_2CO_3 + ZnO$.

VIII. $SrCO_3$

IX. $H_3BO_3 + SrCO_3$

X. $Li_2CO_3 + SrCO_3$

XI. $H_3BO_3 + Li_2CO_3 + SrCO_3$

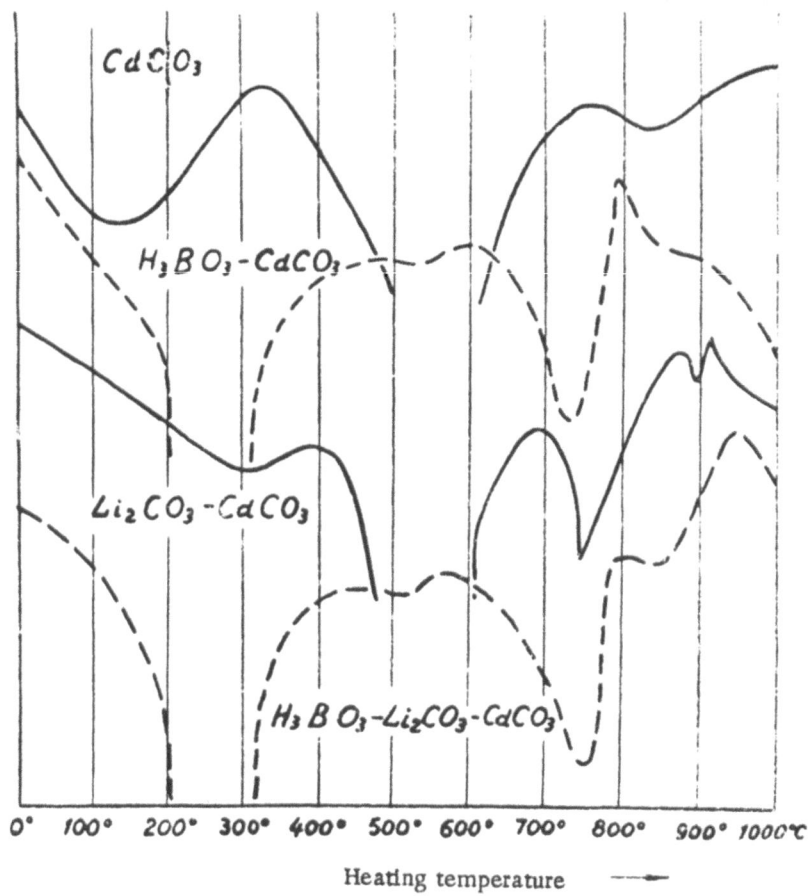

Fig. 27. Thermograms of $CdCO_3$, $CdCO_3 + H_3BO_3$ (Li_2CO_3) and $CdCO_3 + Li_2CO_3 +$ $+ H_3BO_3$.

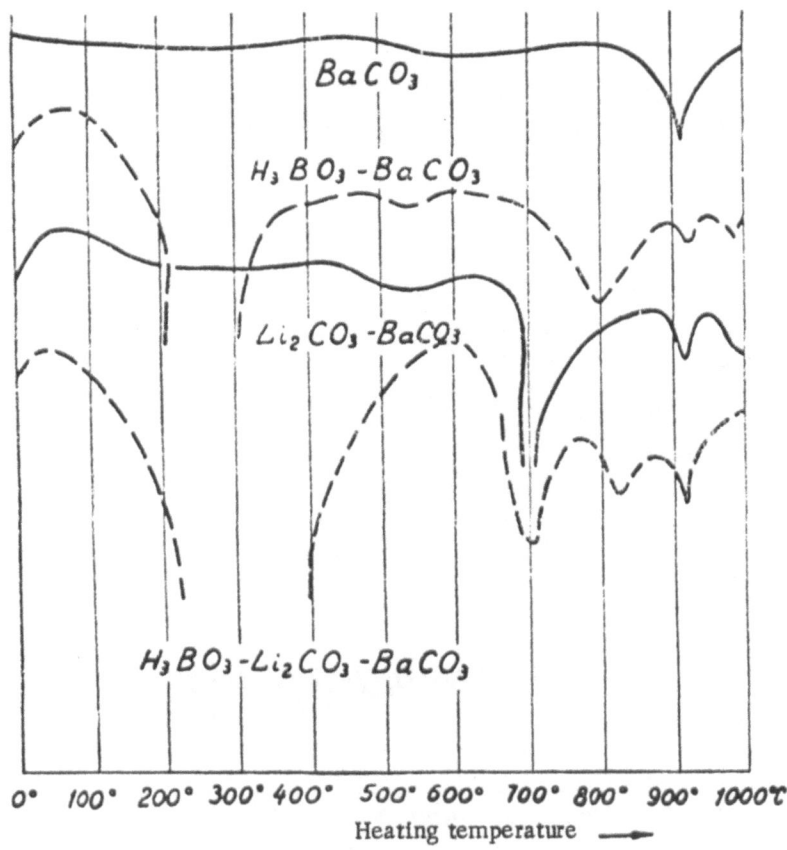

Fig. 28. Thermograms of BaCO₃, BaCO₃ + H₃BO₃ (Li₂CO₃) and BaCO₃ +
+ Li₂CO₃ + H₃BO₃.

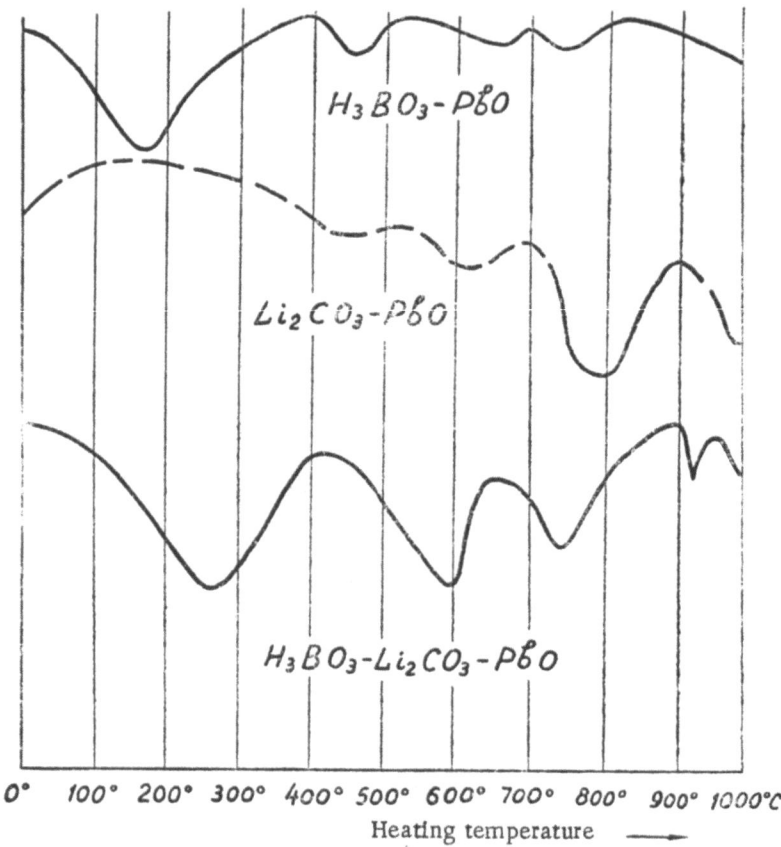

Fig. 29. Thermograms of PbO + H₃BO₃ (Li₂CO₃) and PbO + Li₂CO₃ + H₃BO₃.

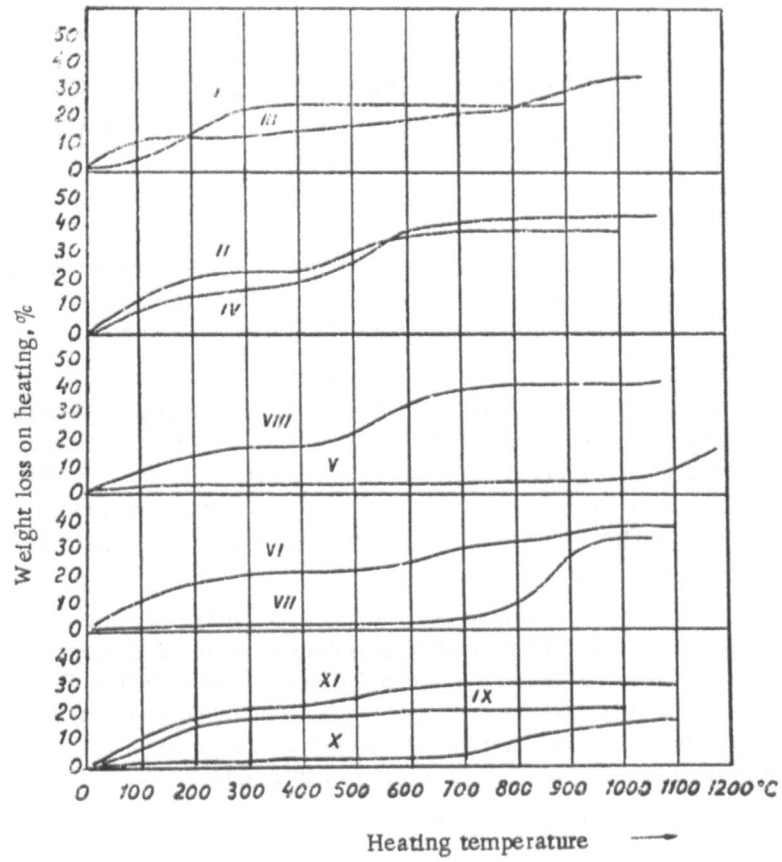

Fig. 30. Weight-loss curves on heating of the following components and their binary and ternary mixtures: CdO_3 ($BaCO_3$, PbO), Li_2CO_3, and H_3BO_3.

I. $CdCO_3$

II. $H_3BO_3 + Li_2CO_3 + CdCO_3$

III. $Li_2CO_3 + CdCO_3$

IV. $H_3BO_3 + CdCO_3$

V. $BaCO_3$

VI. $H_3BO_3 + BaCO_3$

VII. $Li_2CO_3 + BaCO_3$

VIII. $H_3BO_3 + Li_2CO_3 + BaCO_3$

IX. $H_3BO_3 + PbO$

X. $Li_2CO_3 + PbO$

XI. $H_3BO_3 + Li_2CO_3 + PbO$

CRYSTALLIZATION OF BORATE GLASSES AND THE COMPOSITION OF THE CRYSTALLIZATION PRODUCTS

The mineralogy and crystallochemistry of natural and artificial borates have not been studied sufficiently. The crystallization of pure borate (nonsilicate) glasses and also the composition and optical constants of the minerals resulting from their crystallization have not been studied at all.

The glass-formation diagrams (Figs. 1-9) show the regions of glass crystallization on cooling. For a more complete and detailed study of the crystallization products, a series of glass compositions were forcibly crystallized. Crystallization and the crystallization products of the glasses were investigated most completely in the systems; $B_2O_3-Li_2O-BeO$, $B_2O_3 - Li_2O - MgO$, $B_2O_3 - BeO - MgO$, $B_2O_3 - Li_2O - BeO - MgO$, and $B_2O_3 - Li_2O - BeO - MgO + SiO_2$ (Al_2O_3 and ZrO_2).

By treating the crystallized-glass composition with hot hydrochloric acid, in the case of the $B_2O_3 - Li_2O - BeO$ system it was possible to isolate a new mineral, namely, a beryllium borate with the composition $3BeO \cdot B_2O_3$, which has not been described in the literature previously. The mineral was of the rhombic system, biaxial and negative and had the refractive indices Ng = 1.626; Nm = 1.611-1.613; Np = 1.566-1.567; Ng $-$ Np = 0.060. According to A. F. Kapustinskii's formula, the energy constant of this mineral U = Σ ($\kappa_a - \kappa_x)^2$ = 16178 kcal and the charge field of force $\frac{z}{a^2}$ Be = 55.3 $\frac{v}{cm}$. In high-boron glasses (more than 70%) it was found that free B_2O_3 separated and this was rapidly hydrated on the surface.

It was impossible to isolate lithium borates due to their considerable solubility. According to literature data, with more than 30 mol. % alkalies in the glass, the crystallization products contain crystals of the $MeBO_2$ type.

Crystals of $3BeO \cdot B_2O_3$ gave to a glass a dense, porcelain-white appearance and lithium borates and hydrated boric oxides produced an opalescent glass.

Only the mineral $MgO \cdot B_2O_3$ and $2MgO \cdot B_2O_3$ were found among the crystallization products in the $B_2O_3 - Li_2O - MgO$ system. This is explained by the fact that all melting of the mixtures and also subsequent thermal treatment of the glasses were carried out at a temperature of 1200° or below.

The mineral $3MgO \cdot B_2O_3$ was found only in the crystallization products of compositions with a high MgO or MgO + BeO content, melting at temperatures above 1300°. This again confirms the fact that crystallization and the composition of the crystallization products are affected by the composition and ratio of the oxides and also the previous thermal treatment of the glass. Compounds formed during glass formation, especially those showing exothermal effects during their formation (ordered sections), were not totally destroyed after the completion of glass formation and were subsequently detected in the crystallization products of the glasses and also in the physical and physicochemical investigation of the properties of the latter.

In contrast to BeO, MgO somewhat decreased the tendency of glasses to crystallize and the glass-formation regions were found to be greater than in the beryllium system. In this respect, the role of magnesium borates was found to be similar to that of lithium borates. In this system also, the separation of free B_2O_3 on crystallization was observed in high-boron compositions.

The optical constants of magnesium borates have been described in the literature and therefore we present here only their energy constants according to A. F. Kapustinskii's formula: $MgO \cdot B_2O_3 = 5730$ kcal; $2 MgO \cdot B_2O_3 = 9448$ kcal; $3MgO \cdot B_2O_3 = 14090$ kcal.

The strength of the field of force of the cation Mg^{2+}, which has a larger radius than Be^{2+} and a smaller field of force, is $\frac{z}{a^2}Mg = 31.5 \frac{v}{cm}$.

All the glasses crystallized on cooling in the $B_2O_3 - BeO - MgO$ system (alkali-free compositions). The composition of the crystallization products was the same as in the system described above and was determined by the quantitative predominance of BeO or MgO in the composition of the glass mixture and also by the total B_2O_3 content and the melting conditions. The character of the crystallization varied. Glasses with a ratio of oxides $\frac{BeO + MgO}{B_2O_3} \leq 0.33$

with a predominance of BeO and a low MgO content gave clearly expressed crystals when the melt was cooled, and glasses with a ratio $\frac{BeO + MgO}{B_2O_3} \geq 0.33$ with a predominant MgO content in the range 5-15 weight% gave glasses of the emulsion-like type.

This confirms what has been said on the effect of magnesium borates on crystallization and also the fact that the roles of Be^{2+} and Mg^{2+} in the formation of the structure of boron glass are not always identical.

The $B_2O_3 - Li_2O - BeO - MgO$ System

When BeO and MgO were simultaneously present in the glass, the glass-formation region was increased more than when either of them was present alone. The composition of the crystallization products in glasses which crystallized on cooling and also in glasses which were forcibly crystallized was determined by the composition and the ratio of the oxides $\frac{Li_2O + MeO}{B_2O_3}$,

the quantitative amount of each of the oxides Li_2O, BeO and MgO present, and also the previous thermal treatment of the glass. In this system also, glasses of an emulsion-like type were formed, though mainly by low-alkali (up to 2.5% of Li_2O) and high-boron (70-90% of B_2O_3) compositions.

An investigation was made of the effect of adding up to 10 parts by weight of SiO_2, Al_2O_3 or $SiO_2 + Al_2O_3$ to glasses of this composition on their crystallization and the composition of the crystallization products. This hardly changed the general result obtained with glasses without these additives. The addition of SiO_2 and especially Al_2O_3 reduced the crystallization region only insignificantly and beryllium borates predominated in the crystallization products of low-alkali compositions and magnesium borates separated only in high-magnesium and high-alkali compositions where the MgO + Li_2O content was greater than that of BeO. Small amounts of Al_2O_3 (up to 5 parts by weight) had hardly any effect on the process and the character of crystallization of the glasses.

TABLE 3

Composition and Characteristics of Crystallization Products of Glasses of the $B_2O_3 - BeO - Li_2O$ System

Sample No.	Chemical composition of starting glasses			Characteristics of crystallization products
	B_2O_3	BeO	Li_2O	
104	82.5	12.5	5.0	Radial clusters of needle-like crystals with the composition $3BeO \cdot B_2O_3$ and point separations of others. The optical constants of one part of the crystals were similar to $3BeO \cdot B_2O_3$, indicating a polymorphous modification of them. The other part evidently consisted of more complex crystals of the $aB_2O_3 \cdot bBeO \cdot cLi_2O$ type. For the optical characteristics of the needle-like crystals see Sample No. 115. Direct extinction and negative birefringence. Biaxial. Refractive indices: Ng = 1.626; Nm = 1.611; Np = 1.566, i.e., a mineral with the composition $3BeO \cdot B_2O_3$ (Figs. 33, 34, and 39).
105	80.0	12.5	7.5	Long platelets and needles of the mineral and also microlites. Interference colors both bright-and gray and yellow. Direct extinction. Main zone positive (Figs. 35 and 38). Only one crystalline phase found; refractive indices: (Ng 1.625; Nm 1.610; Np 1.565, i.e., a mineral with the composition $3BeO \cdot B_2O_3$.
115	77.5	12.5	10.0	Coarse, irregular, prismatic crystals and also coarse, needle-like crystals and solid mats of fine needles. Main zone positive. Interference colors from light gray to bright of the second and third orders. Extinction was direct relative to the crystallographic directions and oblique relative to cleavage cracks. The mineral was biaxial and negative. Refractive indices: Ng = 1.625; Nm = 1.610; Np = 1.566. i.e., a mineral with the composition $3BeO \cdot B_2O_3$ (Fig. 31). Point separations of another phase were also detected between the crystals of this solid phase (see Sample No. 104).

TABLE 3 (continued)

Sample No.	Chemical composition of starting glasses			Characteristics of crystallization products
	B_2O_3	BeO	Li_2O	
107	85.0	10.0	5.0	The crystals were both irregular prisms and needles in radial clusters and agglomerations. The interference colors were from weak gray and yellow to bright of the second and third orders. Direct extinction. Biaxial. Negative birefringence. Refractive indices: $Ng = 1.625$; $Nm = 1.610$; $Np = 1.566$, i.e., a mineral with the composition $3BeO \cdot B_2O_3$ (Fig. 40).
108	82.5	10.0	7.5	Long tablets and needles and many cross sections with bright interference colors. In optical constants the minerals were similar to those described (Fig. 41).
113	85.0	7.5	7.5	Long irregular platelets and also needle-like crystals. The interference colors were light gray and yellow in some grains and bright in others. Main zone positive. Biaxial and negative. Angle of optical axes $2V = 58°20'$. Refractive indices: $Ng = 1.626$; $Nm = 1.611$; $Np = 1.566-1.567$, i.e., a mineral with the composition $3BeO \cdot B_2O_3$. Birefringence = 0.055. The platelets and tablets had the same refractive index as boric acid (Fig. 42).
106	87.5	10.0	2.5	Radial clusters of crystals. Point crystals of another substance between the rays. The coarser crystals had weak interference — light gray and yellow tones. The mineral was biaxial and the main zone positive. Negative birefringence. The phenomenon of direct extinction was observed. Polysynthetic twins were present. Refractive indices: $Ng = 1.626$; $Nm = 1.610$; $Np = 1.565$, i.e., a mineral with the composition $3BeO \cdot B_2O_3$ (Fig. 45).
118	80.0	2.5	17.5	Coarse, partially scaly crystals without regular faces. Interference colors of the third order. Uniaxial. Negative birefringence = 0.059. Main zone positive. Refractive indices of part of the crystals: $Ng = 1.625$; $Np = 1.563$. Similarity of the optical constants to those of the mineral $3BeO \cdot B_2O_3$ indicates that another modification of this mineral could have been present in this case (Fig. 43).

TABLE 3 (continued)

Sample No.	Chemical composition of starting glasses			Characteristics of crystallization products
	B₂O₃	BeO	Li₂O	
117	92,5	5.0	2.5	See description of Sample No. 118 (Fig. 44)
111	77,5	15.0	7.5	The same
112	60,0	20.0	20.0	Minerals analogous to previous samples, see Sample No. 113, etc. (Figs. 35-37)
114	72,5	15.0	12.5	The same
116	80,0	10.0	10.0	The same

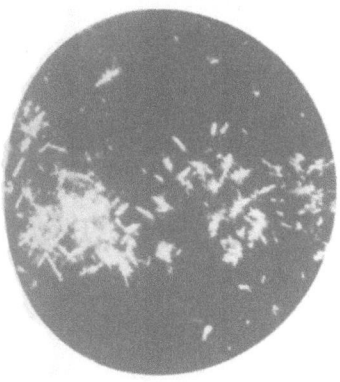

Fig. 31. Sample No. 115/11.
Nicols ×, magnification 80×.

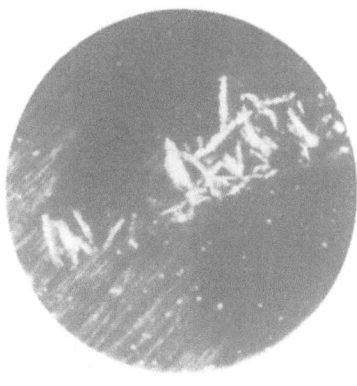

Fig. 32. Sample No. 105a.
Nicols ×, magification 80×.

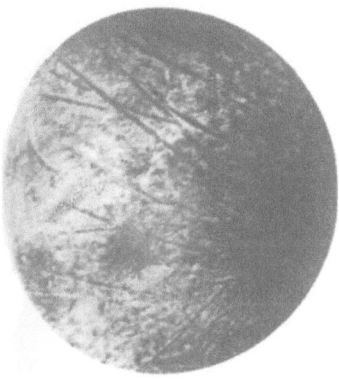

Fig. 33. **Sample No. 104a.**
Nicols ×, **magnification 80×.**

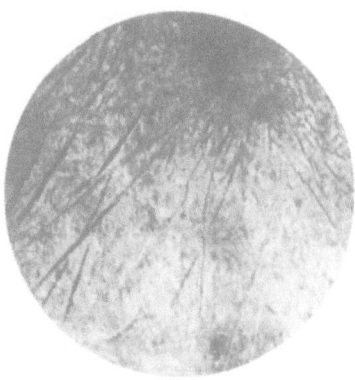

Fig. 34. Sample No. 104b.
Nicols ‖, magnification 80×.

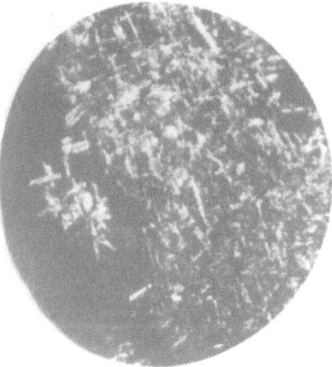

Fig. 35. Sample No. 111d.
Nicols ×, magnification 80×.

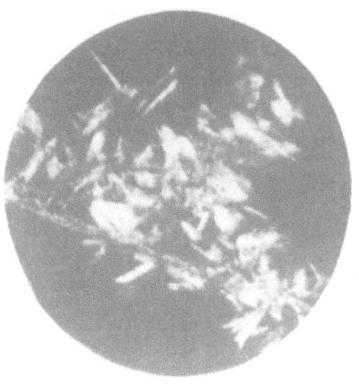

Fig. 36. Sample No. 112/1.
Nicols ×, magnification 80×.

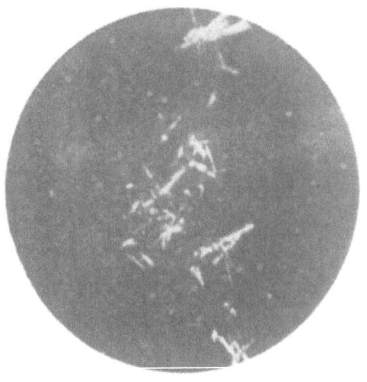

Fig. 37. Sample No. 114/1.
Nicols ×, magnification 80×.

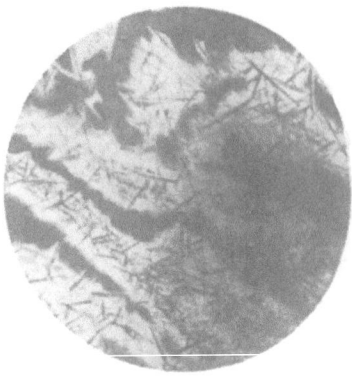

Fig. 38. Sample No. 105 /10.
Nicols ∥, magnification 80×.

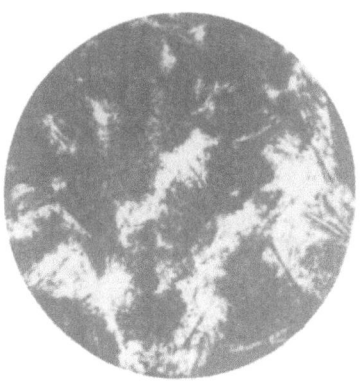

Fig. 39. Sample No. 104e.
Nicols ∥, magnification 62×.

Fig. 40. Sample No. 107/17.
Nicols ∥, magnification 62×.

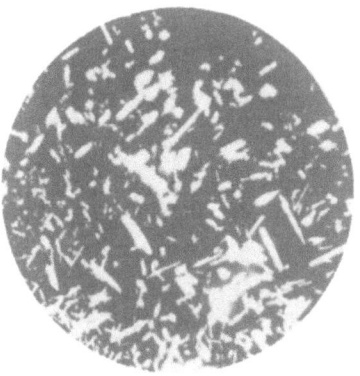

Fig. 41. Sample No. 108a.
Nicols ×, magnification 62×.

Fig. 42. Sample No. 113a.
Nicols ×, magnification 80×.

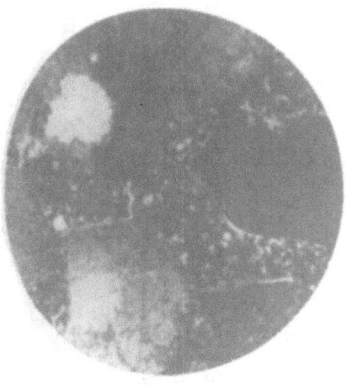

Fig. 43. Sample No. 118/47.
Nicols ×, magnification 62×.

Fig. 44. Sample No. 117/31.
Nicols ‖, magnification 62×.

Fig. 45. Sample No. 106/16.
Nicols ‖, magnification 62×.

Fig. 46. Sample No. 69.
Nicols ×, magnification 80×.

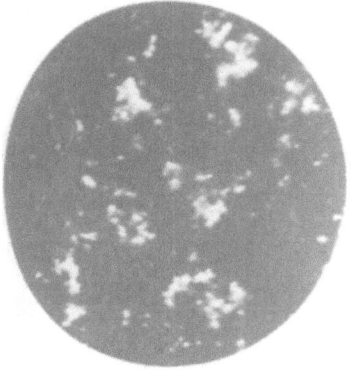

Fig. 47. Sample No. 74.
Nicols ×, magnification 80×.

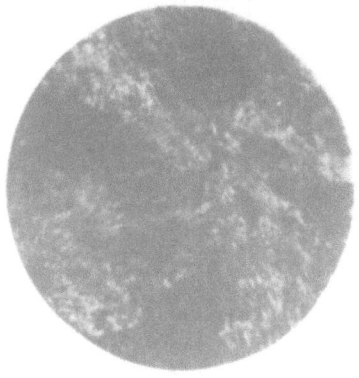

Fig. 48. Sample No. B-64.
Nicols ×, magnification 80×.

TABLE 4

Sample No.	Chemical composition of starting glasses			Characteristics of crystallization products
	B_2O_3	MgO	Li_2O	
63	75.0	15.0	10.0	Clusters of fine plate-like crystals with clearly expressed lengthwise cleavage, high birefringence, and some times direct, and at others oblique, extinction (C : Np up to 25°). The mineral was biaxial with a high value of 2V close to 90° and negative. The refraction was high: Ng ≦ 1.670; Np ~ 1.580; Ng − Np = 0.09. Complex fan-like or spherulite forms of crystals were sometimes formed. The mineral had the composition $2MgO \cdot B_2O_3$ (Fig. 49).
64	70.0	20.0	10.0	Coarse crystals with high interference of an optically biaxial, negative mineral with high birefringence and refraction. The mineral had the composition $2MgO \cdot B_2O_3$ (Fig. 48).
65	70.0	15.0	15.0	There were both coarse plate-like and very elongated crystals with voluminous polysynthetic twinning and lengthwise cleavage. High birefringence and refraction. Np ≦ 1.590. The mineral had the composition $2MgO \cdot B_2O_3$ (Figs. 50 and 46).
69	89.5	8.5	2.0	
66	90.5	7.0	2.5	Together with crystals corresponding to the composition $2MgO \cdot B_2O_3$, there were very fine point crystals with very low birefringence and a refraction slightly less than that of the glass. They were mainly boric acids (Figs. 52, 53, and 47).
74	94.0	1.0	5.0	
72	96.5	1.0	2.5	The whole glass was thickly covered with crystals of a leaf-like and sometimes dendritic form with birefringence up to orange tones of the first order and low refraction: Nm ~ 1.471. These formations were mainly boric acids.
68	94.5	3.0	2.5	
71	93.5	4.0	2.5	

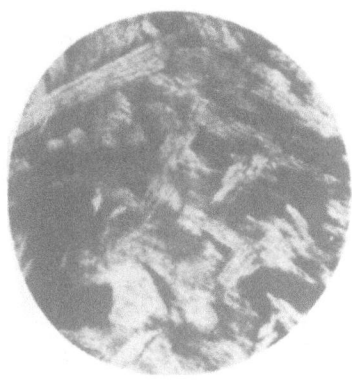

Fig. 49. Sample No. A-63.
Nicols ×, magnification 80×.

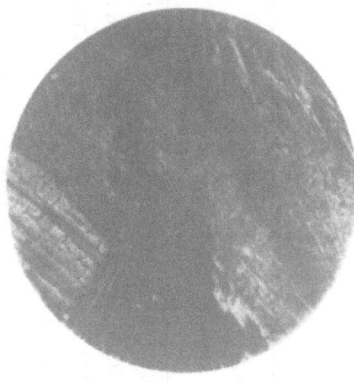

Fig. 50. Sample No. B-65.
Nicols ×, magnification 80×.

Fig. 51. Sample No. 129.
Nicols ×, magnification 80×.

Fig. 52. Sample No. 66b.
Nicols ×, magnification 80×.

Fig. 53. Sample No. 66c.
Nicols ×, magnification 80×.

Fig. 54. Sample No. 99.
Nicols ×, magnification 80×.

TABLE 5

Composition and Characteristics of Crystallization Products of Glasses of the $B_2O_3 - BeO - MgO$ System

Sample No.	Chemical composition of starting glasses			Characteristics of crystallization products
	B_2O_3	BeO	MgO	
124	90.0	5.0	5.0	Fine plate-like crystals with clearly expressed cleavage 010 with a predominance of direct extinction and high birefringence and refraction. Numerous polysynthetic twins. Refractive index Ng $\leq$ 1.627. The mineral had the composition $3BeO \cdot B_2O_3$ (Figs. 55, 57 and 59).
129	89.0	10.0	1.0	Fine plate-like crystals with direct extinction, positive elongation and high birefringence and refraction. Nm $\geq$ 1.607. The mineral had the composition $3BeO \cdot B_2O_3$ (Fig. 51).
128	95.0	2.5	2.5	Solid crystallization. Three crystalline phases (Fig. 58): 1. Scales with low refraction, which changed rapidly during determination, and birefringence up to white tones of the first order — boric acids. 2. Groups of point crystals with a refraction lower than that of the glass. 3. Platelets of a mineral with high birefringence and refraction — magnesium borates.
121 125	85.0 80.0	10.0 10.0	5.0 10.0	Separate crystalline platelets with average birefringence similar to boric acid.
122	75.0	10.0	15.0	Dense, interlaced mass of crystals of leaf-like form with low birefringence — boric acids.

TABLE 5 (continued)

Sample No.	Chemical composition of starting glasses			Characteristics of crystallization products
	B_2O_3	BeO	MgO	
127	90.0	—	10.0	Two crystalline phases: 1. Fine scaly crystals predominated — boric acids. 2. Sections with a large amount of platelets of a biaxial, negative mineral with a high value for 2V. The optical constants were similar to those of the mineral with the composition $2MgO \cdot B_2O_3$ (Fig. 56).
130	90.0	10.0	—	Two crystalline phases (Fig. 60): 1. Fine, scaly, irregular agglomerations of crystals with a low refraction and an average birefringence — boric acids. 2. Fine, elongated platelets, almost always polysynthetically twinned with direct extinction, positive elongation and high birefringence and refraction. Minerals of the $3BeO \cdot B_2O_3$ type.

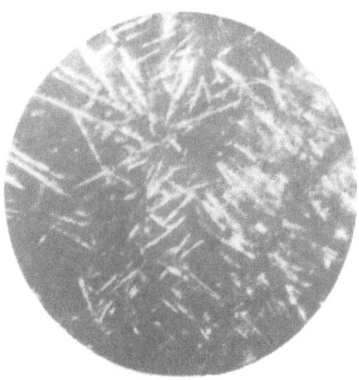

Fig. 55. Sample No. 4j-124.
Nicols ×, magnification 80×.

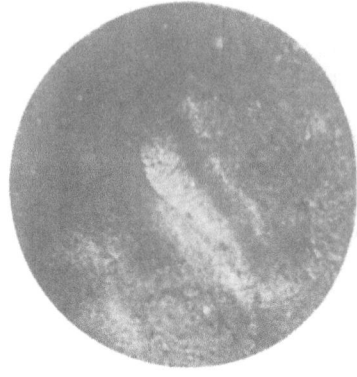

Fig. 56. Sample No. 2h-127.
Nicols ×, magnification 80×.

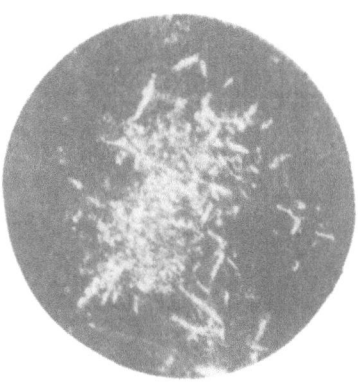

Fig. 57. Sample No. 124a.
Nicols ×, magnification 80×.

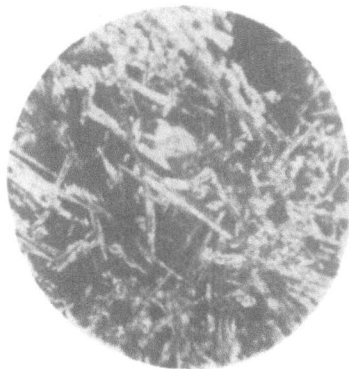

Fig. 58. Sample No. 3i-128.
Nicols ×, magnification 80×.

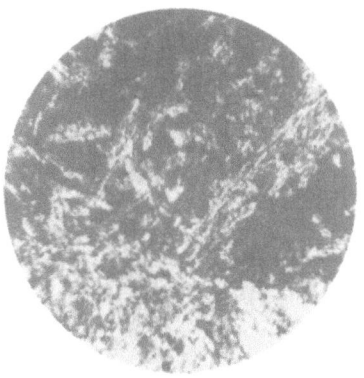

Fig. 59. Sample No. 124b.
Nicols ×, magnification 80×.

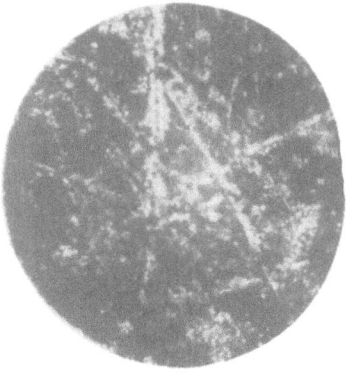

Fig. 60. Sample No. 1g-130.
Nicols ×, magnification 80×.

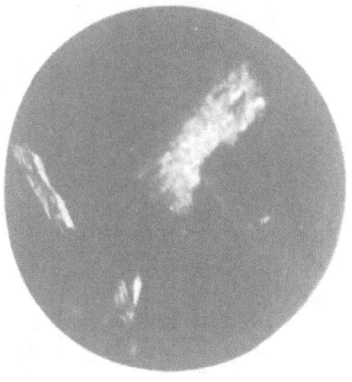

Fig. 61. Sample No. 50b.
Nicols ×, magnification 80×.

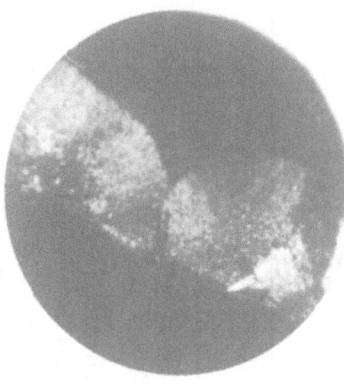

Fig. 62. Sample No. 480-47.
Nicols ×, magnification 80×.

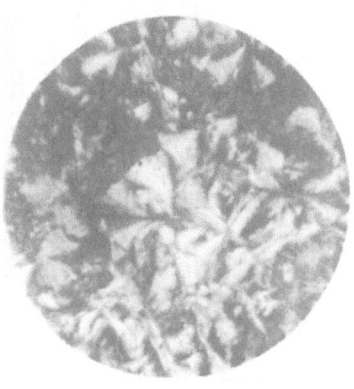

Fig. 63. Sample No. 52.
Nicols ×, magnification 80×.

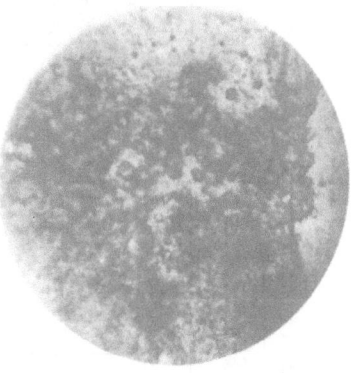

Fig. 64. Sample No. 16-7.
Nicols ×, magnification 80×.

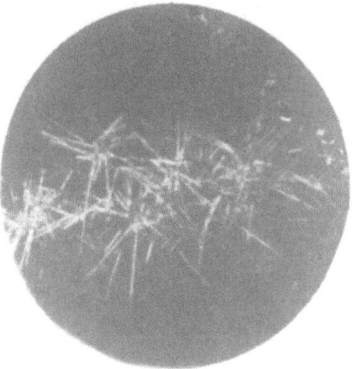

Fig. 65. Sample No. 16-58.
Nicols ×, magnification 80×.

Fig. 66. Sample No. 270-53.
Nicols ×, magnification 80×.

TABLE 6

Composition and Characteristics of Crystallization Products of Glasses of the $B_2O_3 - Li_2O - BeO - MgO$ System

Sample No.	Chemical composition of starting glasses					Characteristics of crystallization products
	B_2O_3	BeO	MgO	Li_2O	$Al_2O_3 + SiO_2$	
50	57.5	12.5	15.0	15.0	$2Al_2O_3$	Fine, plate-like crystals. For optical characteristics see Sample No. 63. Mineral $2MgO \cdot B_2O_3$ (Fig. 61).
47	67.5	5.0	15.0	2.5	$10Al_2O_3 + 10SiO_2$	The whole of the glass was uniformly covered with very fine point crystals with a refraction somewhat less than that of the glass. Sections were observed with spherulites or single, coarser platelets of a biaxial, negative mineral with an average value of 2V and high birefringence and refraction. The mineral had the composition $3BeO \cdot B_2O_3$ (Fig. 62).
52	70.0	2.5	10.0	2.5	$5Al_2O_3 + 10SiO_2$	See description of Sample No. 50 (Fig. 63).
7	94.0	2.5	1.0	2.5	$2Al_2O_3$	Boric acids, see description of Sample No. 3 (Fig. 64).
58	75.0	2.5	15.0	7.5	—	Fine plate-like crystals, always twinned along the length, birefringence up to blue tones of the first order. Direct extinction predominated, but there was sometimes weak oblique extinction. Positive elongation. High refraction: Ng = 1.658. Mineral $MgO \cdot B_2O_3$ (Fig. 65).
40	57.5	15.0	15.0	12.5	—	Spherulite agglomerations consisting of petal-shaped fibrous forms. Polysynthetic twinning was noticeable. The extinction of the fibers was sometimes direct and sometimes oblique. Refractive indices Ng = 1.670, Np = 1.652, Ng − Np = 0.02. The mineral had the composition $3MgO \cdot B_2O_3$. Crystals with the composition $3BeO \cdot B_2O_3$ were found (Fig. 68).

TABLE 6 (continued)

Sample No.	Chemical composition of starting glasses					Characteristics of crystallization products
	B_2O_3	BeO	MgO	Li_2O	$Al_2O_3 + SiO_2$	
42	55.0	15.0	15.0	15.0	$5Al_2O_3$	Fine, fibrous crystals with high birefringence, biaxial, negative and with a high value of 2V. Np ≥ 1.588, i.e., a mineral of the $2MgO \cdot B_2O_3$ type (Fig. 76).
43	55.0	15.0	15.0	15.0	$10Al_2O_3$	The same as in Sample No. 42. Separate sections of hexagonal prisms with high refraction and average birefringence — BeO crystals (Fig. 67).
53	65.0	7.5	15.0	12.5	—	Fine plate-like, sometimes fibrous crystals. Extinction sometimes direct and sometimes oblique. High birefringence and refraction. The mineral was biaxial and negative with 2V ~90°, i.e., $2MgO \cdot B_2O_3$ (Fig. 66).
99	62.5	10.0	15.0	12.5	—	Fine plate-like, sometimes fibrous crystals. Extinction sometimes direct and sometimes oblique. High birefringence and refraction. Biaxial and negative with 2V ~ 90°, i.e., $2MgO \cdot B_2O_3$ (Fig. 54).
48	72.5	5.0	5.0	2.5	$5Al_2O_3 + 10SiO_2$	Three crystalline phases: 1. Clusters of fine platelets with high birefringence, biaxial, negative mineral with average value of 2V and Nm ≤ 1.609, i.e., $3BeO \cdot B_2O_3$. 2. Clusters of fine fibrous crystals with refraction close to 1.600. Apparently a mixture of $3BeO \cdot B_2O_3$ with some other component. 3. Leaf-like crystals with constants characteristic of B_2O_3 hydrates (Fig. 75).

TABLE 6 (continued)

Sample No.	Chemical composition of starting glasses					Characteristics of crystallization products
	B_2O_3	BeO	MgO	Li_2O	$Al_2O_3 + SiO_2$	
59	65.0	10.0	15.0	10.0	—	Plate-like and complex fibrous forms of a monoclinic, biaxial, negative mineral with $2V \sim 90°$ and high birefringence and refraction i.e., the mineral $2MgO \cdot B_2O_3$. There were also minerals of the $3BeO \cdot B_2O_3$ type.
3	95.5	2.5	1.0	1.0	$10Al_2O_3$	Two types of crystals: 1. Platelets with broken edges with direct extinction, high birefringence and indistinctly expressed twinning of a uniaxial, negative mineral with low refraction — B_2O_3 hydrates. 2. Very fine, indeterminate scales of crystals with a higher refraction than that of the glass (Fig. 74).
15	67.5	5.0	15.0	12.5	—	Platelets and sometimes sheaf-like, fibrous crystals. Extinction sometimes direct and at others oblique. $Ng \leq 1.670$, $Np \sim 1.580$, i.e., the minerals $2MgO \cdot B_2O_3$ (Figs. 69 and 70).
55	70.0	2.5	15.0	12.5	—	Elongated platelike crystals with complete lengthwise cleavage and with high birefringence and refraction. The mineral was biaxial and negative and had a high value of $2V$, i.e., $2MgO \cdot B_2O_3$ (Fig. 71).
56	72.5	2.5	15.0	10.0	—	The same (Fig. 72).

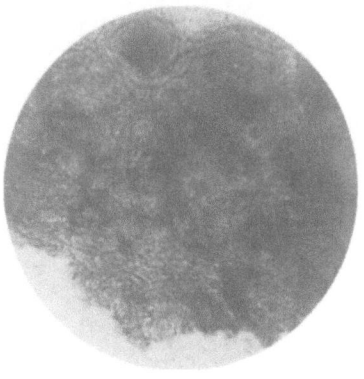

Fig. 67. Sample No. 555-43.
Nicols ×, magnification 80×.

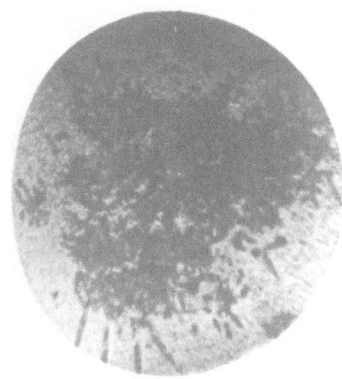

Fig. 68. Sample No. 540-40.
Nicols ×, magnification 80×.

Fig. 69. Sample No. 15a.
Nicols ×, magnification 80×.

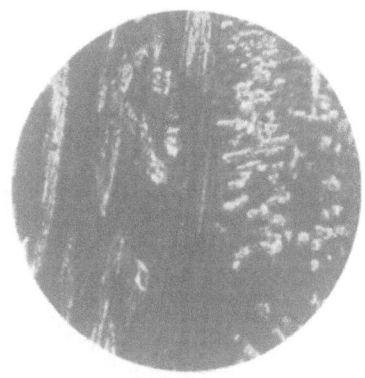

Fig. 70. Sample No. 180-15.
Nicols ×, magnification 80×.

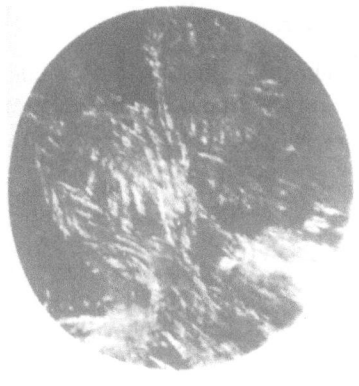

Fig. 71. Sample No. 90-55.
Nicols ×, magnification 80×.

Fig. 72. Sample No. 90-56.
Nicols ×, magnification 80×.

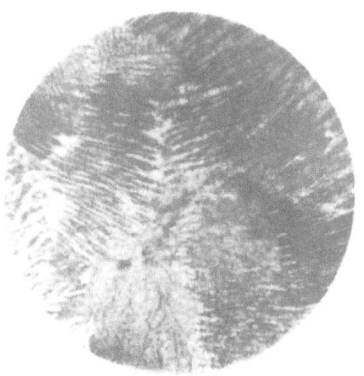

Fig. 73. Sample No. 345-59.
Nicols ×, magnification 80×.

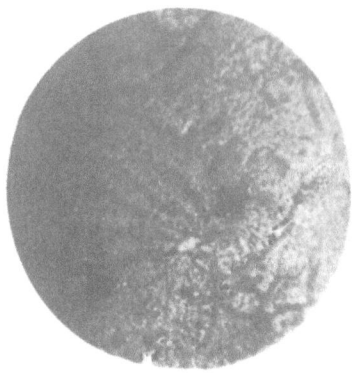

Fig. 74. Sample No. 3.
Nicols ×, magnification 80×.

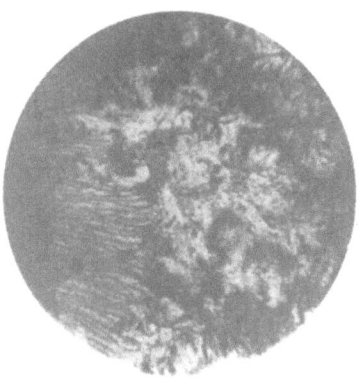

Fig. 75. Sample No. 48.
Nicols ×, magnification 80×.

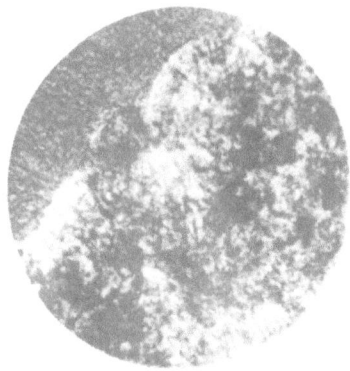

Fig. 76. Sample No. 42.
Nicols ×, magnification 80×.

It was established that more complex borates of the type $aBeO \cdot bLi_2O \cdot cB_2O_3$ were also formed in all glasses, particularly beryllium glasses. It was impossible to determine their exact optical constants by a microscopic method due to their point character.

The compositions of a series of crystallized glasses of the $B_2O_3 - Li_2O-BeO$, $B_2O_3-Li_2O--MgO$, $B_2O_3 - BeO - MgO$ and $B_2O_3 - Li_2O - BeO - MgO$ systems and the characteristics of the crystallization products are presented in Tables 3-6 and photoreproductions of microsections are given in Figs. 31-76.

At the present time, the presence of crystals and bubbles in glasses is by no means always a fault in them.

It is known that the structure of crystallized glass is characterized by not only the structure of the monocrystals, but also their dimensions and the conditions of their coalescence and mutual disposition.

Thus, the results of investigating the crystallization not only solves problems of the structure of borate glasses, but also serve as a basis for the preparation of crystallizing glasses with a given structure, in particular, glasses with a high mechanical strength.

PHYSICOCHEMICAL PROPERTIES OF BORATE GLASSES

An investigation was made of the physicochemical properties of glasses of all the systems presented to determine their fields of application and to establish the character of the change in the properties of the glasses with a continuous change in their composition.

The three following main conflicting points of view on the character of the changes in the properties of simple glasses with a continuous change in their compositions were established at the All-Union Conference on the Structure of Glass [13]:

1. The properties undergo discrete changes at compositions corresponding to definite chemical compounds (K. S. Evstrop'ev).

2. The properties undergo discrete changes at compositions corresponding to phase boundaries on the phase diagram (K. G. Kumanin, O. K. Botvinkin, and L. I. Demkina).

3. The properties change continously with a change in composition (A. A. Appen).

The properties of silicate and borosilicate glasses, including the behavior of boric oxide in silicate glasses have been studied relatively fully. There has been very little investigation of the properties of pure borate glasses and the effect of boric oxide on the properties of these glasses. A large number of experiments and the analysis of their results are necessary to determine the rules and character of the change in the properties of glasses, in this case borate glasses, with a change in their composition.

DENSITY

Literature data on the factors affecting the density and other properties of glass do not always agree. A number of investigators reported [92, 107, 153] that a change in the coordination number of the elements has the greatest effect on the density of glass. According to Kühne [153], the ionic radii affect the density of glass to a lesser extent.

The effect of B_2O_3 on the density of glass is far from clear. Appen, Abe, Grjotheim and others [92, 105, 107, 153] reported that the introduction of a definite amount of alkali into borate glasses consolidates their structure and increases the density. According to Warren and others, the greatest density of glass corresponds to tetracoordination of boron and according to Grjotheim, to tricoordination.

The values of the density of different oxides at the same molecular-percent content are proportional to their molecular weights (Tables 7 and 8). The problem of the effect of Li_2O on the density of glass is more complex. Some investigators [112] reported that Li_2O has a specific consolidating effect on the structure of glass and that Li_2O facilitates the preparation of more-dense glasses.

The density of the glass was investigated pycnometrically in xylene and toluene. The absolute densities of the glasses investigated ranged from 1.9 to 4.35 g/cm^3 and are presented in Tables 9 and 13-20 (appendix) and the ternary density diagrams are given in Fig. 77-85.

As a rule, B_2O_3 reduced and Li_2O increased the density of the glass. The oxides MeO increased the density of the glass in relation to their molecular weights. An increase in the

67

TABLE 7

Range of Changes in the Densities of Glasses from the Systems Investigated in Relation to the Nature of the Oxide MeO, the Change in the Amount of the Oxides Present and the Ratio $\dfrac{m Li_2O + n MeO}{B_2O_3}$

		BeO	MgO	BeO+MgO	CaO	ZnO	SrO	CdO	BaO	PbO
D, g/cm³	from	2,00	2,05	2,00	1,90	2,14	2,10	2,16	2,25	2,12
	to	2,28	2,38	2,48	2,47	2,80	3,10	3,60	3,91	4,35
$\dfrac{m Li_2O + n MeO}{B_2O_3}$	from	0,30	0,22	0,22	0,20	0,18	0,17	0,16	0,16	0,15
	to	1,00	0,55	1—1,4	0,67	0,67	0,67	0,67	0,67	0,67

TABLE 8

Effect of the Oxide MeO on the Change in Density of Glass with the Composition B_2O_3 — 60 mol.%; Li_2O — 5 mol.%; MeO — 35 mol.%

	BeO	MgO	CaO	ZnO	SrO	CdO	BaO	PbO
D of glass ($D_{BeO} = 1,0$)	1,0	1,05	1,08	1,13	1,36	1,56	1,72	1,90

radii of the cations and also the interatomic distances of the oxides had a definite effect on the change in the molecular volume of the glass.

The character of the change in the density curves varied. With an increase in the MeO + + Li_2O content of the glass up to a definite limit [~ 40 mol.% of BeO + Li_2O; ~30 mol.% of MgO (CaO, ZnO, SrO, CdO, BaO) + Li_2O and ~20 mol.% of PbO + Li_2O]. the density curves of the glasses rise extremely steeply. On reaching the given amount of the oxides nMeO + $m Li_2O$ and the corresponding ratio $\dfrac{m Li_2O + n MeO}{B_2O_3}$, the density curves of the glasses show a sharp break and subsequently become comparatively flat (Figs. 110-117).

Thus, the introduction of the oxides nMeO + $m Li_2O$ into the glass up to the limits indicated first produced a marked consolidation of the structure of the glasses and then the increase in density slowed down. This characterizes the state of the structure and, consequently, the ratio of tetra- and tricoordinate boron in the glass and the degree and nature of borate formation and determines the character of the change in density with a change in the composition of the oxides.

It is quite evident that the character of the change in the density of the glasses depends not so much on the simple quantitative change in the composition of the oxides as on the change in their quantitative ratio to each other and the ratio $\dfrac{m Li_2O + n MeO}{B_2O_3}$, i.e., on the coordination state of the boron, the nature and composition of the borates formed and the degree of binding of the boron in the glass. No specific consolidating action of Li_2O in the given glasses was found.

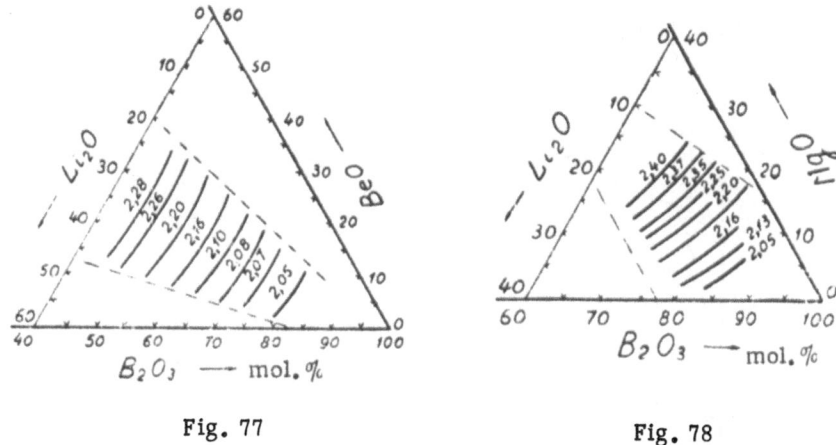

Fig. 77   Fig. 78

Fig. 77. Density diagram of glasses with the composition $B_2O_3 - Li_2O -$ $-BeO$ (g/cm^3).

Fig. 78. Density diagram of glasses with the composition $B_2O_3 - Li_2O -$ $- MgO$ (g/cm^3).

It should also be noted that in the $B_2O_3 - Li_2O - MgO$ system, at different ratios $\dfrac{nB_2O_3}{mLi_2O}$ in the glass there were extrema in the change in density with the change in MgO content (Fig. 79). At $\dfrac{B_2O_3}{Li_2O} \sim 16$, extrema were obtained at MgO contents within the limits of 1-1.5 molecular parts. At $\dfrac{B_2O_3}{Li_2O} \sim 8$, there were extrema at MgO contents of 0.5 and 1.5 molecular parts; at $\dfrac{B_2O_3}{Li_2O} \sim 5$, there were extrema at MgO contents of 0.2 and 0.3 molecular parts.

This anomalous change in density may be explained by the different degree of binding of MgO and B_2O_3 in borates (1:1, 2:1 and 3:1), the corresponding decrease in the amount of unbound B_2O_3 present and also the effect of the quantitative amount of MgO and also Li_2O on the change in the coordination state of the boron in the glass and its general structure.

Such extrema were not found in the $B_2O_3 - Li_2O - BeO$ system.

Such sharply expressed extrema were not found in the $B_2O_3 - Li_2O - BeO - MgO$ quaternary system, i.e., with BeO and MgO present simultaneously, and this was evidently due to the fact that BeO was also present.

This again confirms what has been said above on the different effect of the given oxides on the structure of the glass and on its properties, in this case, on the density.

REFRACTIVE INDICES

In textbooks on crystallochemistry [26, 37, 154] it is stated that refraction depends mainly on the anions and increases with an increase in size and a decrease in the solidity of the electron shell, that the ionic refraction is an additive property of the polarization of the given ions [26, 153] and that the optical properties are generally insensitive to structure and depend only on the electronic structure and the type of the chemical bonds of the oxides [26]. According to

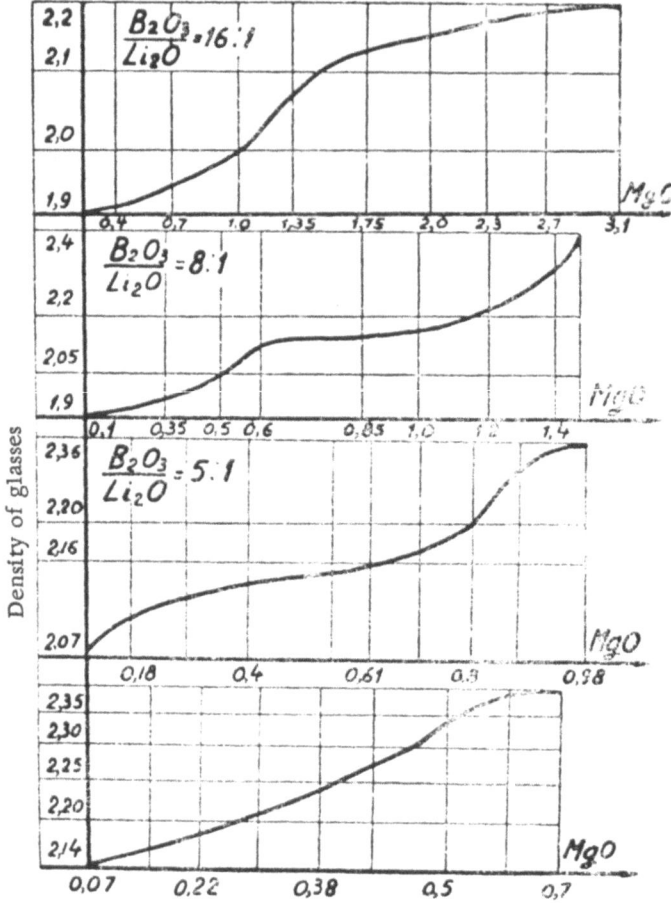

Fig. 79. Anomalous changes in the density of glasses in the B_2O_3 —
$Li_2O - MgO$ system at various $\dfrac{B_2O_3}{Li_2O}$ ratios and MgO contents.

Kühne [153], the specific refraction is also affected by an increase in valence and, to a lesser extent, an increase in the radius of an ion and its coordination number.

The relation between the refractive index of a glass and its density is frequently expressed by the following empirical formulas [174]:

$$\frac{n-1}{\rho} = C \quad \text{(Gladstone and Dale's formula)}$$

$$\frac{n^2-1}{(n+0.4)\rho} = C \quad \text{(Eichman's formula)}$$

$$\frac{n^2-1}{\rho} = C \quad \text{(Newton and Drude's formula)}$$

$$\frac{n^2-1}{(n^2+2)\rho} = C \quad \text{(the most frequently used formula of Lorentz and Lorenz).}$$

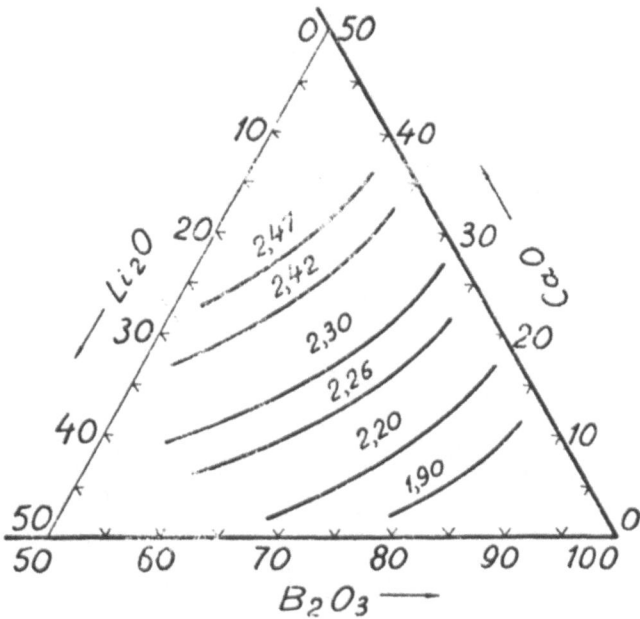

Fig. 80. Density diagram of glasses with the composition $B_2O_3 - Li_2O - CaO$ (g/cm^3).

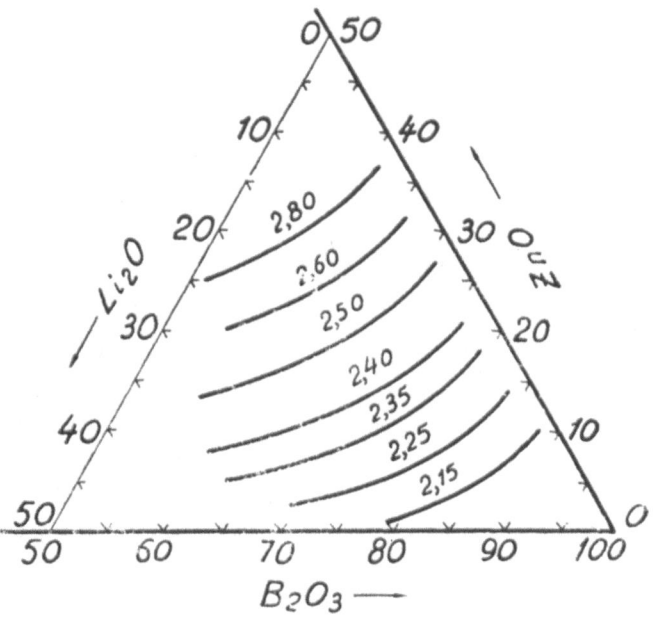

Fig. 81. Density diagram of glasses with the composition $B_2O_3 - Li_2O - ZnO$ (g/cm^3).

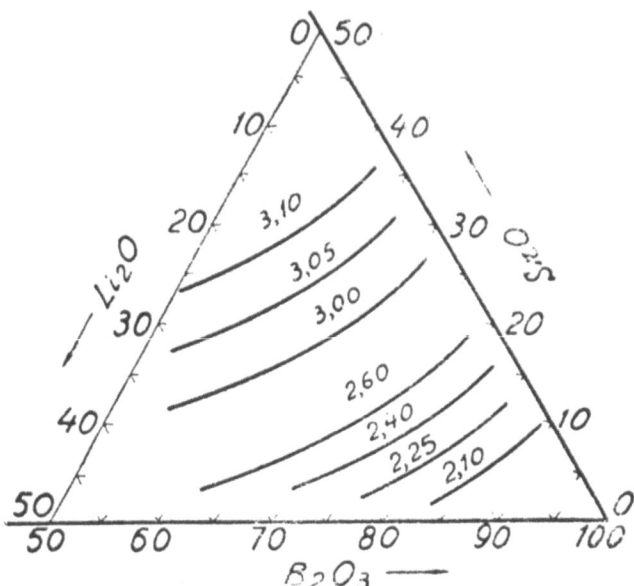

Fig. 82. Density diagram of glasses with the composition
$B_2O_3 - Li_2O - SrO$ (g/cm^3).

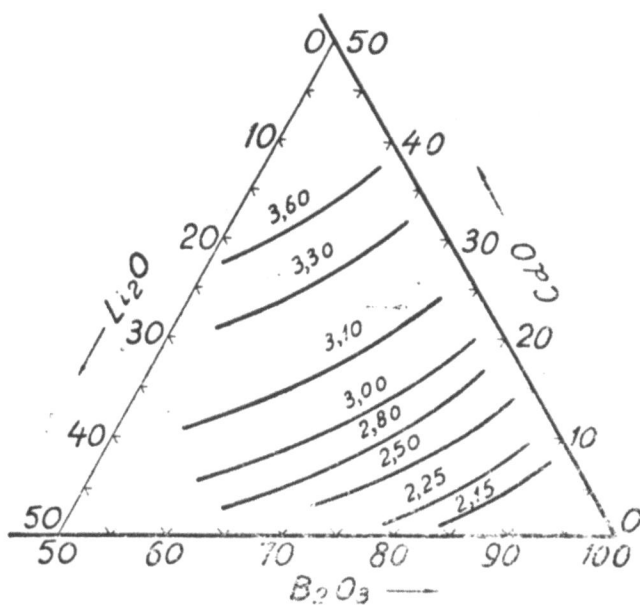

Fig. 83. Density diagram of glasses with the composition
$B_2O_3 - Li_2O - CdO$ (g/cm^3).

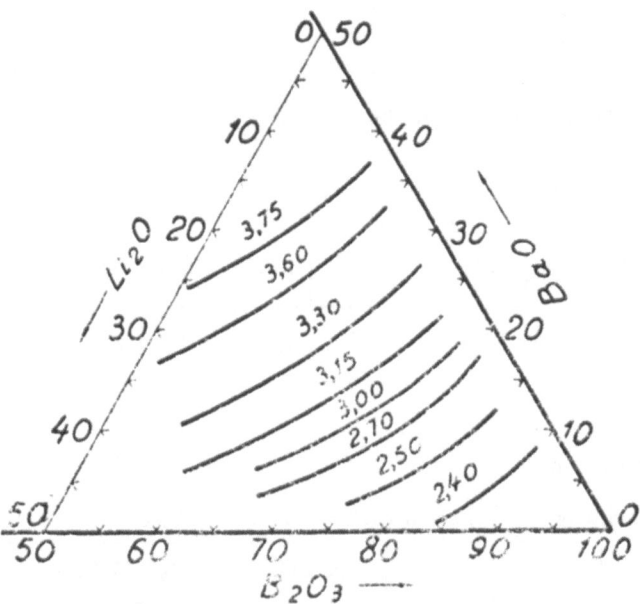

Fig. 84. Density diagram of glasses with the composition
$B_2O_3 - Li_2O - BaO$ (g/cm³).

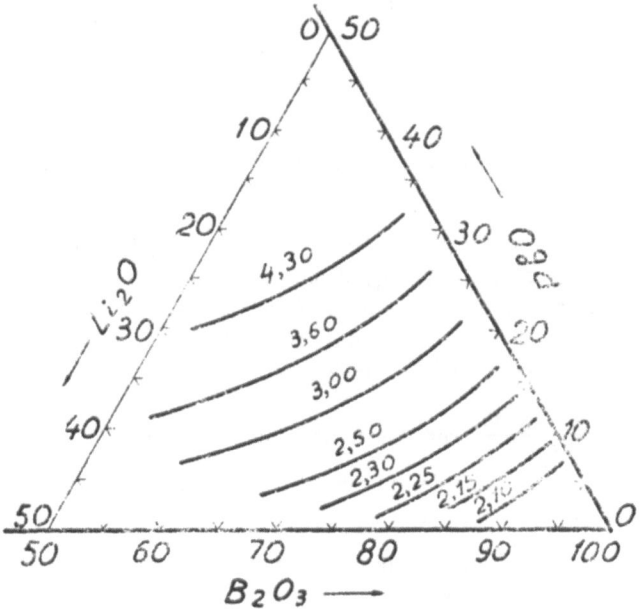

Fig. 85. Density diagram of glasses with the composition
$B_2O_3 - Li_2O - PbO$ (g/cm³).

TABLE 9

Range of Changes in the Refractive Indices of the Glasses Investigated in Relation to the Nature of the Oxide MeO, the Change in the Amount of the Oxides Present and the Ratio $\dfrac{Li_2O + MeO}{B_2O_3}$

		BeO	MgO	BeO+MgO	CaO	ZnO	SrO	CdO	BaO	PbO
N_D	from	1,520	1,515	1,507	1,515	1,507	1,508	1,513	1,503	1,515
	to	1,574	1,577	1,577	1,599	1,596	1,596	1,653	1,615	1,715

TABLE 10

Compositions of Glasses in mol. % and Their N_D

B_2O_3	Li_2O	MeO	N_D of glasses with MeO						
			MgO	CaO	ZnO	SrO	CdO	BaO	PbO
80	5	15	1,522	1,540	1,541	1,541	1,559	1,552	1,606
70	5	25	—	1,572	1,576	1,580	1,604	1,579	1,666
60	15	25	—	1,599	1,586	1,594	1,630	1,608	1,672
60	5	35	—	1,592	1,596	1,596	1,653	1,615	1,715

Here $\underline{n}$ is the refractive index, ρ the density and C a constant.

The calculated average value of C for the glasses investigated is 0.135 (for the Lorentz-Lorenz formula).

The refractive indices of the glasses were determined by the immersion method by means of a polarizing microscope. The N_D of the immersion liquids was checked before their use with a precision refractometer. The values of the refractive indices of the glasses investigated varied from 1.5 to 1.75 and are presented in Tables 9 and 13 to 20 (appendix) and the refractive-index diagrams are given in Figs. 86-93.

The values of $\dfrac{Li_2O + MeO}{B_2O_3}$ were the same as in Table 7 (p. 75).

The effect of the different oxides on the refractive index is shown in Figs. 110-117. As a rule, B_2O_3 considerably reduced N_D and Li_2O increased it.

The relative effect of different oxides MeO on the N_D of glasses can be seen from a comparison of the N_D values for successive groups of glasses.

The comparative effect of the oxides MeO on the increase in N_D of glasses is approximately in the following order: BeO, MgO, CaO, ZnO, SrO, CdO, BaO and PbO. In all cases,

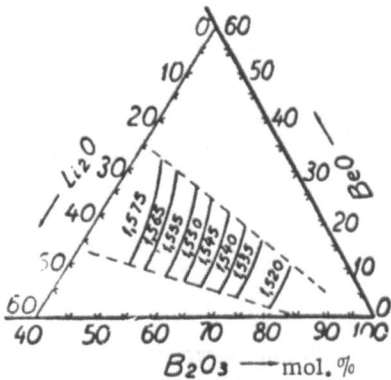

Fig. 86. Refractive-index diagram of glasses with the composition $B_2O_3 - Li_2O - BeO$.

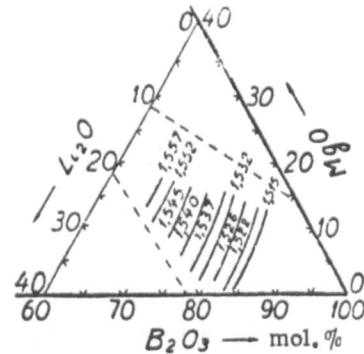

Fig. 87. Refractive-index diagram of glasses with the composition $B_2O_3 - Li_2O - MgO$.

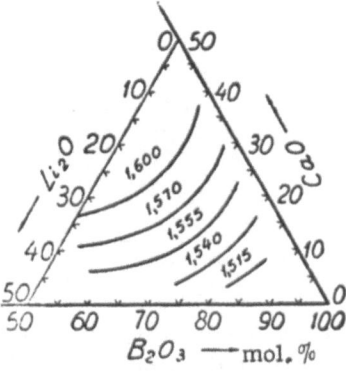

Fig. 88. Refractive-index diagram of glasses with the composition $B_2O_3 - Li_2O - CaO$.

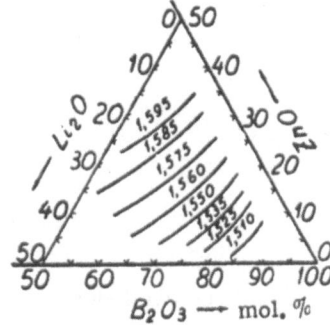

Fig. 89. Refractive-index diagram of glasses with the composition $B_2O_3 - Li_2O - ZnO$.

cadmium oxide increased the N_D of the glasses more than BaO. Like the density curves, the N_D curves of the glasses rose very steeply with an increase in the $mLi_2O + nMeO$ content of the glass up to definite ratios of $\dfrac{mLi_2O + nMeO}{B_2O_3}$, which are presented above. The N_D curves were almost parallel to the density curves up to their breaks. When the given ratios of $\dfrac{mLi_2O + nMeO}{B_2O_3}$ were exceeded, breaks were also observed in the N_D curves, but these were not so sharply expressed as on the density curves.

Thus, the character of the change in N_D of the glasses depends on the change in the ratio of $\dfrac{mLi_2O + nMeO}{B_2O_3}$ and, consequently, on the structure of the glass, as these are interrelated.

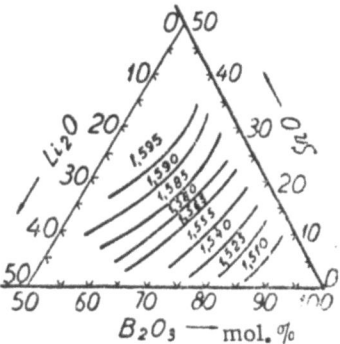

Fig. 90. Refractive-index diagram
of glasses with the composition
$B_2O_3 - Li_2O - SrO$.

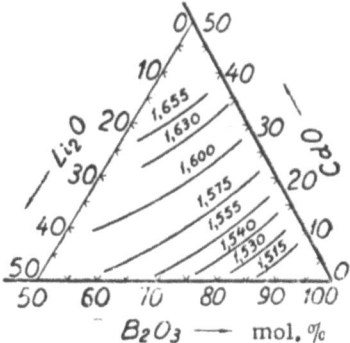

Fig. 91. Refractive-index diagram
of glasses with the composition
$B_2O_3 - Li_2O - CdO$.

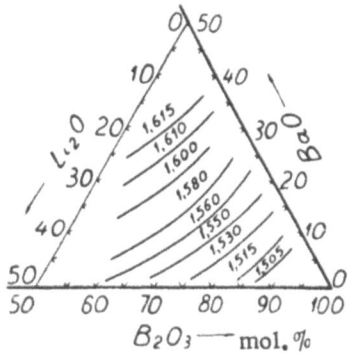

Fig. 92. Refractive-index diagram
of glasses with the composition
$B_2O_3 - Li_2O - BaO$.

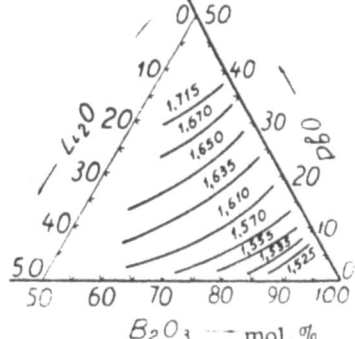

Fig. 93. Refractive-index diagram
of glasses with the composition
$B_2O_3 - Li_2O - PbO$.

The results of the investigation do not confirm the ideas expressed in crystallo-chemistry textbooks, according to which the optical properties of glass are insensitive to structure.

CHEMICAL STABILITY

As is known, the chemical stability of glasses is closely connected with their structure. The chemical stability of borate glasses is undoubtedly connected with the ratio of $BO_4 : BO_3$ in the glass and the degree of binding of B_2O_3 in borates. Many investigators consider that the chemical stability of a glass is maximal if the amount of oxides Me_2O introduced into the glass is sufficient to convert 1/5 of the boron into the corresponding coordination state (BO_4 or BO_3), which is most stable for boron. As has already been stated, some investigators [90-92] relate this to the formation XY_4 structural groups ($X = BO_4$ and $Y = BO_3$). A greater amount of MeO and Me_2O promotes disintegration of the structure $= B-O^-...^-O-B =$ and, consequently, the chemical stability of the glass must fall.

It is reported that lithium glasses are more stable chemically than analogous sodium glasses [110-115], while the effect of the oxides MeO varies in relation to the amount of them in the glass.

76

Unfortunately, the literature data on the chemical stability of glasses and the effect of separate components on their stability cannot be compared as the investigators used different procedures (powder method, swelling, boiling glass powder in water for various times and keeping samples in a reagent medium for various times). Some data, for example, those of Moore [155] and conclusions based on tests on from 3 to 10 glass compositions are not very convincing.

We tested the water resistance of glasses from all the systems investigated since it is of exceptionally great importance in their use. It should be noted that the determination of the water resistance of borate glasses is a very difficult problem since a number of glasses have a considerable solubility ("leachability") and some are completely destroyed by water.

The products from water "leaching" consisted not only of alkalis, as with silicate glasses, but largely of Li_2O and B_2O_3 and sometimes contained the oxides MeO also.

The residue after "leaching," especially of glasses with a low MeO content, was frequently a gelatinous mass — a gel; flakes of it even passed through the walls of platinum gauze baskets in which the water resistance was determined. Also, the solutions were often extremely difficult to filter.

We used a powder method to investigate the water resistance of glasses. We did not consider the swelling method worthwhile in this case for determining the behavior of glasses during their use under the most drastic conditions since this method makes no allowance for the actual state of the surface of glassware under working, heating, and other conditions, nor does it distinguish the roles of the separate components or the effect of structure on the chemical stability of the glass. The most objective (but also the most drastic) method of testing the chemical stability of these glasses is a determination of the solubility of glass powder with a fresh fracture surface and a considerable surface area of the grains in the appropriate reagent liquid. It was established that even chemically unstable glasses undergo little external corrosion during prolonged (many months) storage indoors, while they rapidly corroded outdoors under drastic conditions and were found to be unsuitable for use. For this reason, until recently borate glasses (especially for "windows" in x-ray tubes) have been used with a lacquer to protect their surface from corrosion.

We determined the water resistance of the glasses in the following way. A 2 g sample of glass powder (weighed on an analytical balance) with a grain size of 0.4 mm (< 0.5 and > 0.3 mm), which had been dried in a drying cupboard, was placed in a quartz flask with 50 ml of distilled water and boiled on a water bath for 1 hour (and for a number of glasses, for 5 hours), while the level of the water in the flask was kept constant by means of reflux condensers. Some samples of glass powder were placed in a platinum-gauze basket (not in a quartz flask), which was then placed in a flask with water. After the boiling period, the flask was rapidly cooled under running water and the solution filtered through an ashless filter (where necessary, a Schott filter was used to collect the precipitate). The residue on the filter was washed 5-6 times and left overnight for complete filtration and to detect any precipitate of gel-like flakes passing through the filter.

The filtrate was added portionwise to a platinum or silver dish, which had been weighed on an analytical balance, evaporated on a water bath, dried in an electric muffle furnace at approximately 300°, and then cooled in a desiccator and weighed. As a result, we obtained the weight of glass which had dissolved.

The accuracy of the results was checked by determining the weight of the residual powder on the filter or in the platinum basket, which was carefully collected and dried to constant weight. In a number of cases the "leaching" products or the residue were analyzed chemically to establish the nature of the "leachability" of glasses of the given type.

Since the glasses do not fit into the normally accepted hydrolysis classification of solubility, we adopted an arbitrary classification which determined the possibility of using the glasses with or without protection of their surfaces with a lacquer.

The arbitrary hydrolysis classification of the glasses investigated was as follows: class I — solubility of glass powder in water less than 2 wt. % of the sample, class II — up to 10 wt. %, class III up to 20 wt. %, class IV — up to 50 wt. % and class V — above 50 wt. %.

The solubility of the glass powder when the latter was boiled in water for 1-5 hours (Tables 9 and 13-20 in the appendix and Figs. 94-101) varied from 96 to 0.2%, which made it possible to establish especially clearly the factors affecting the change in water resistance of the glasses.

The effect of different oxides on the water resistance of glass is shown in Figs. 110-117. The greatest solubility in water was shown by glass with BeO or MgO and the least by glass containing BeO and MgO simultaneously and also glass with CaO, ZnO, and CdO.

The over-all solubility of the glass in water was found to be minimal when the whole of the boron was bound in borates and maximal when lithium borates predominated in the glass and when part of the B_2O_3 was unbound.

The formation of any borates, even lithium borates, sharply reduced the solubility of the glass in water.

As can be seen from Table 11, the solubility of the glasses in water decreased proportionally to the increase in the ratio $\dfrac{mLi_2O + nMeO}{B_2O_3}$, especially as a result of an increase in the content of the oxides MeO and, consequently, due to a predominant increase in the contents of MeO borates and also proportionally to the increase in density of the glass. Glasses with a high content of MeO borates and with a high density were similar in solubility to normal industrial silicate and borosilicate glasses.

Such a sharp fall in the solubility of the glass in water from 96.5 to 0.2% cannot be explained only by the simple quantitative change in the composition of the oxides; it is also connected with the change in structure of the glass, the increase in the density of its packing and the formation of definite borates, the corresponding change in the coordination state of the boron, and the reduction of the amount of unbound B_2O_3 to a minimum.

The solubility curves of the glasses were analogous to the curves for the change in density of glasses of the same compositions.

The ratio O/B in the glasses in all the series increased to 1.78-1.93 and even to 2 and more (see Tables 9 and 13-20 in the appendix). However, this did not lead to disintegration of the structure of the glass and an increase in solubility, but, on the contrary, promoted consolidation of the structure of the glass and a sharp fall in its solubility in water, especially with an increase in the oxygen content due to MeO.

Consequently it was not simple changes in the amount of oxides that determined the degree of solubility of the glass in water, but mainly structural factors: the ratio of the oxides

TABLE 11

Water Solubility of Different Series of Glasses and Factors on Which Their Water Resistance Depends

Compositions of glasses in mol %			% Solubility in water — maximum and minimum	$\dfrac{MeO + Li_2O}{B_2O_3}$	Density of glasses — minimum and maximum
B_2O_3	Li_2O	MeO			
65,0	28,3	BeO— 6,7	93,6	0,5	2,1132
47,5	24,5	BeO— 28,0	36,3	1,1	2,2719
85,6	9,7	MgO— 4,8	96,4	< 0,2	2,0503
75,1	10,1	MgO— 14,8	9,2	0,5	2,2436
82,0	10,4	BeO+MgO—7,7	96,5	< 0,25	2,09
45,3	12,3	BeO+MgO—42,4	0,21	1,25	2,36
83,6	10,7	CaO— 5,7	Gel	< 0,25	1,90
60,0	15,0	CaO— 25,0	1,5	0,65	2,42
85,0	11,0	ZnO— 4,0	52,0	0,2	2,14
60,0	5,0	ZnO— 35,0	1,0	0,67	2,80
85,8	11,0	SrO— 3,2	69,0	0,2	2,10
60,0	5,0	SrO— 35,0	3,0	0,66	3,10
86,2	11,2	CdO— 2,6	78,8	< 0,2	2,16
60,0	5,0	CdO— 35,0	0,3	0,67	3,60
86,8	11,1	BaO— 2,1	87,0	< 0,2	2,45
60,0	5,0	BaO— 35,0	5,0	0,67	3,91
87,3	11,2	PbO— 1,5	87,0	0,15	2,12
60,0	5,0	PbO— 35,0	1,5	0,67	4,33

$\dfrac{mLi_2O + nMeO}{B_2O_3}$, the degree of binding of B_2O_3 in borates, the character of the borates formed, and the corresponding change in the coordination state of the boron in the glass.

THERMAL EXPANSION OF GLASS

Kühne [153] indicated that the ionic radius and also the coordination number had a predominating effect on the increase in the thermal expansion of glass. The greater the inter-atomic distance of the oxide, the greater was the value of $\alpha \cdot 10^{-7}$. The structure with the greatest packing density had the least expansion coefficient. The value of $\alpha \cdot 10^{-7}$ decreased

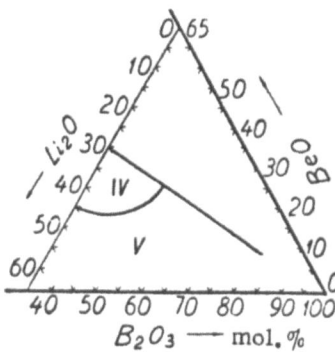

Fig. 94. Water resistance (% solu-
bility) diagram of glasses with the
composition $B_2O_3 - Li_2O - BeO$:
I — solubility up to 2%
II — solubility up to 10%
III — solubility up to 20%
IV — solubility up to 50%
V — solubility above 50%.

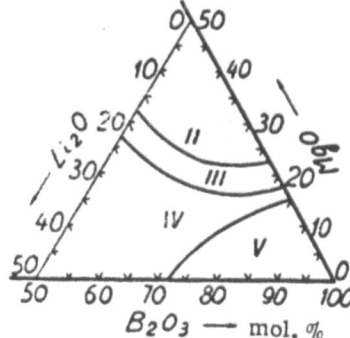

Fig. 95. Water resistance (% solubility) diagram of glasses with the
composition $B_2O_3 - Li_2O - MgO$.

The value of $\alpha \cdot 10^{-7}$ decreased with an increase in the valence of the elements and the
minimum was observed for tetravalent elements.
With a further increase in the valence, the value
of $\alpha \cdot 10^{-7}$ increased.

A number of investigators [73] found that the expansion curve was relatively stable for
lithium glasses at O/B between 1.71 and 1.78. With an O/B ratio between 1.78 and 1.98 more
intensive displacements in the lattice were observed.

An increase in the ratio of O/B in the glass above that necessary for the formation of a
dense structure in the glass, especially increases due to oxygen of "modifier" ions, promoted
disintegration of the structure and, consequently, an increase in the expansion of the glass.
In the general case, any decrease in the regularity of the glass lattice increased the expansion
coefficient. This means that heat supplied was first spent in the glass in ordering the lattice
and then in its direct expansion [73, 177].

The coefficients of thermal expansion were determined on a dilatometer over the temperature range 20-420° for glasses with the composition $B_2O_3 - Li_2O - BeO$ and 20-320° for
glasses of the other systems. The range of the change in the thermal-expansion coefficients
of the glass in relation to the nature of the oxide MeO, the change in the quantitative content
of the oxides and the ratio $\dfrac{mLi_2O + nMeO}{B_2O_3}$ is presented in Table 12.

The values of $\dfrac{mLi_2O + nMeO}{B_2O_3}$ were the same as in Table 7 (p. 75).

The values of the expansion coefficients for glasses of all the systems are presented in
Tables 9, 13-20 and 22 (appendix) and in Figs. 102-109.

The effect of different oxides on the thermal-expansion coefficient of the glass is shown
in Figs. 110-117.

Analysis of the expansion curves of the glasses shows that at the quantitative $Li_2O + MeO$
content and the ratios $\dfrac{Me_2O}{B_2O_3}$ and $\dfrac{Me_2O + MeO}{B_2O_3}$ indicated above (see "density"), the glasses
showed clearly expressed extrema on the thermal-expansion curves, which corresponded

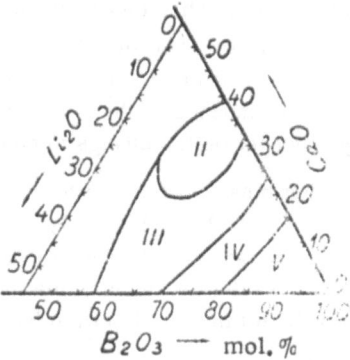

Fig. 96. Water resistance (% solubility) diagram of glasses with the composition $B_2O_3 - Li_2O - CaO$.

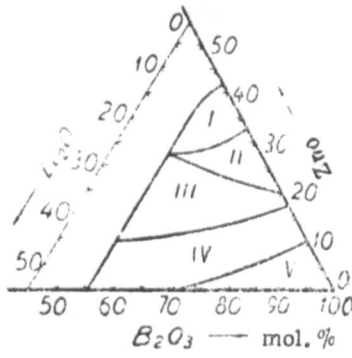

Fig. 97. Water resistance (% solubility) diagram of glasses with the composition $B_2O_3 - Li_2O - ZnO$.

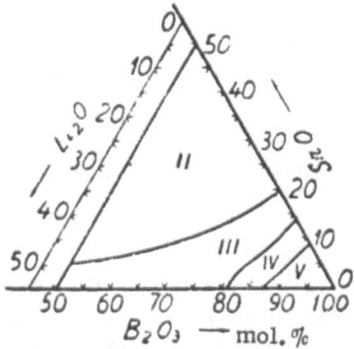

Fig. 98. Water resistance (% solubility) diagram of glasses with the composition $B_2O_3 - Li_2O - SrO$.

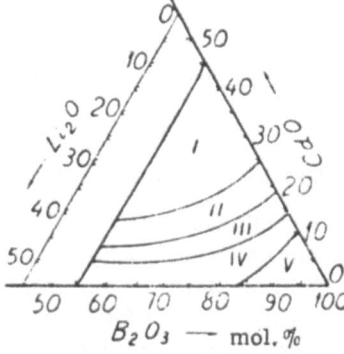

Fig. 99. Water resistance (% solubility) diagram of glasses with the composition $B_2O_3 - Li_2O - CdO$.

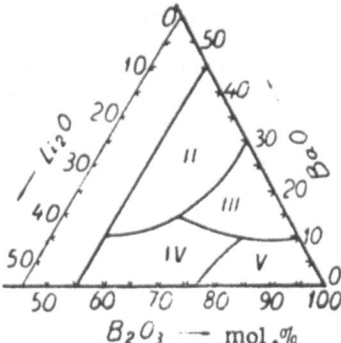

Fig. 100. Water resistance (% solubility) diagram of glasses with the composition $B_2O_3 - Li_2O - BaO$.

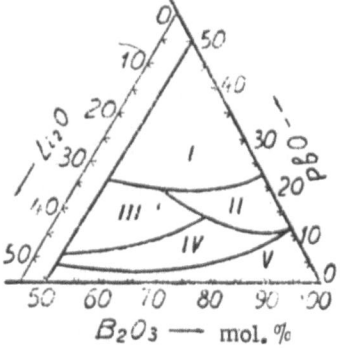

Fig. 101. Water resistance (% solubility) diagram of glasses with the composition $B_2O_3 - Li_2O - PbO$.

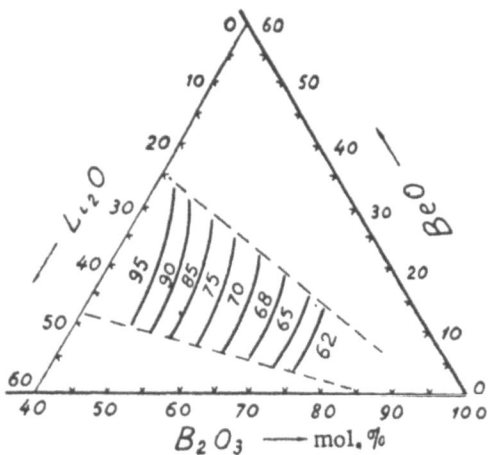

Fig. 102. Thermal expansion ($\alpha \cdot 10^{-7}$) diagram of glasses with the composition $B_2O_3 - Li_2O - BeO$.

basically to the character of the density curves of glasses from the given systems, and this leads to the same conclusions regarding the dependence of the character of property changes on the change in composition.

For binary systems with the composition $Me_2O - B_2O_3$ and $MeO - B_2O_3$, we and other investigators [92] found the formation of definite minima and maxima corresponding to the formation of certain borates.

The character of the change in the thermal expansion curves of borate glasses shows that up to the ratio $\dfrac{Li_2O + MeO}{B_2O_3}$ given above, "compression" of the samples occurs, i.e., heat supplied is spent in consolidating and ordering the structure of the structure of the glass, but with a further change in the ratio $\dfrac{Li_2O + MeO}{B_2O_3}$ there is quite vigorous expansion in the glass and a corresponding sharp rise in the expansion curve.

TRANSPARENCY AND ABSORPTION OF SHORT-WAVE RADIATION

Until recently only glasses with the composition $B_2O_3 - Li_2O - BeO$ have been used as x-ray-transparent materials and these have a transparency coefficient of ~ 0.7 for plates 1.0 cm thick and ~ 0.97 for plates 0.1 cm thick. Correspondingly, the absorption coefficient of these glasses for x-radiation is 0.3 for 1.0 cm plates and 0.03 for 0.1 cm plates. Since the glasses have a considerable solubility, the surface of the glass was covered with a layer of lacquer for protection against corrosion by moisture and other reagents.

Compositions with a high thermal-expansion coefficient of the order of $90 \cdot 10^{-7}$, suitable for sealing to platinum electrodes, were used for sealing to electrovacuum glass. Where molybdenum and tungsten electrodes were used, it was found necessary to employ glasses with $\alpha \cdot 10^{-7} < 50$ and the glasses already applicable required the presence of special glasses for so-called graded seals between "platinum" and "tungsten" ("molybdenum") glasses.

Our investigations considerably extended the possibility of applying any composition of glass with a high water resistance and with a given thermal-expansion coefficient for sealing to glasses and electrodes without graded seals. Some difficulty arises in the working of glass compositions with a low $\alpha \cdot 10^{-7}$ since here it is necessary to introduce considerable amounts of B_2O_3 which increases their tendency to crystallize during prolonged holding of the melt at high temperature and also necessitates rapid working and cooling.

Experimental determination of the x-ray transparency for glasses with the composition $B_2O_3 - Li_2O - BeO$ [126] showed that the atomic or molecular absorption coefficients for x-rays of the separate elements were additive. This permitted simple calculation of the transparency or absorption of x-rays by glasses of various compositions [69-70, 156].

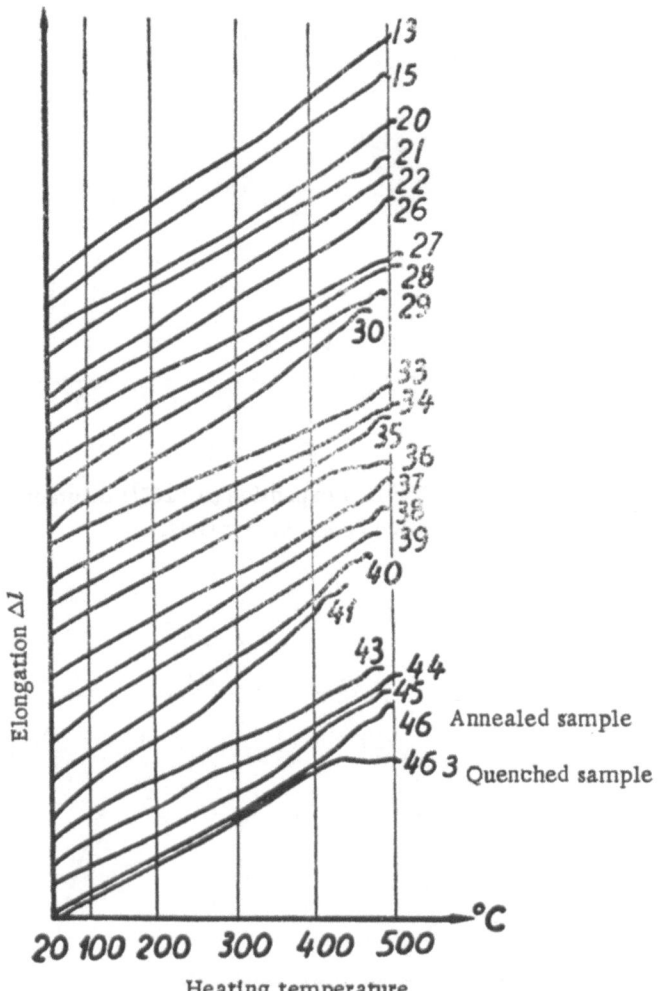

Fig. 103. Character of expansion curves of glasses with
the composition $B_2O_3-Li_2O-BeO$.

The calculation was carried out by means of the formula

$$I=I_0e^{-\mu t} \quad or \quad \frac{I}{I_0}=e^{-\mu t}.$$

At $t = 1$ (1 cm) $\frac{I}{I_0} = e^{-\mu}$.

We have $\frac{I}{I_0} = X$, $X = e^{-\mu}$.

Taking logariphms of both parts of the equation we obtain $\log x = -\mu \log e = -\mu\ 0.4343$,
a polycomponent glass $\log x = -\Sigma\ \mu\ 0.4343 = -D\ \Sigma\ (f\ \omega)\ 0.4343$. Absorption of x-rays by a
glass.

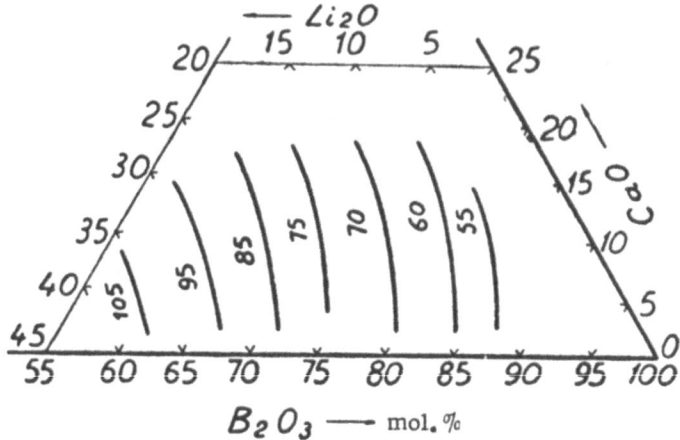

Fig. 104. Thermal expansion ($\alpha \cdot 10^{-7}$) diagram of glasses with the composition $B_2O_3 - Li_2O - CaO$.

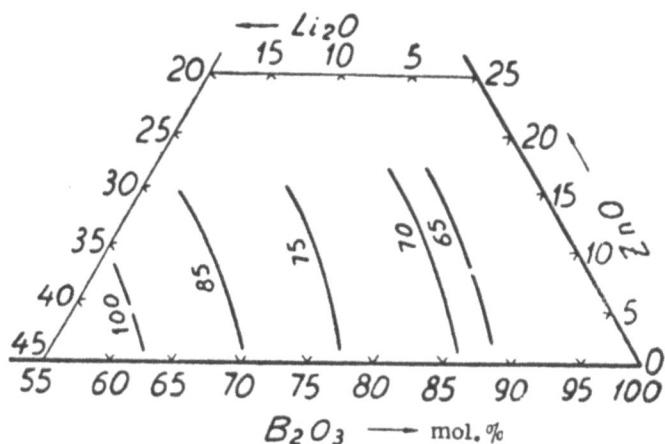

Fig. 105. Thermal expansion ($\alpha \cdot 10^{-7}$) diagram of glasses with the composition $B_2O_3 - Li_2O - ZnO$.

$$A = 1 - \frac{I}{I_0} = 1 - X.$$

Where I_0 is the total intensity of the x-radiation, I is the intensity of this radiation after passing through the glass, t is the thickness of the absorbing or transmitting body in cm, μ is the absorption coefficient $= D \cdot \Sigma (f \cdot \omega)$ cm^{-1}, ω is the mass absorption coefficient, D is the density of the given glass in g/cm^3, and f is the weight content of the oxides, $\Sigma f = 1$.

ωB_2O_3 at λ = 0.1 A = 0.1385,
ωMgO " λ " = 0.1466
ωLi_2O " λ " = 0.1325
ωBeO " λ " = 0.1359

The constants for calculating the x-ray transparency of the glasses are presented in Tables 23 and 24 (appendix).

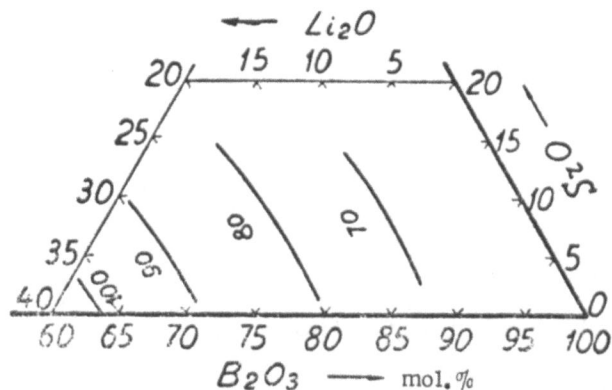

Fig. 106. Thermal expansion ($\alpha \cdot 10^{-7}$) diagram of glasses with the composition $B_2O_3 - Li_2O - SrO$.

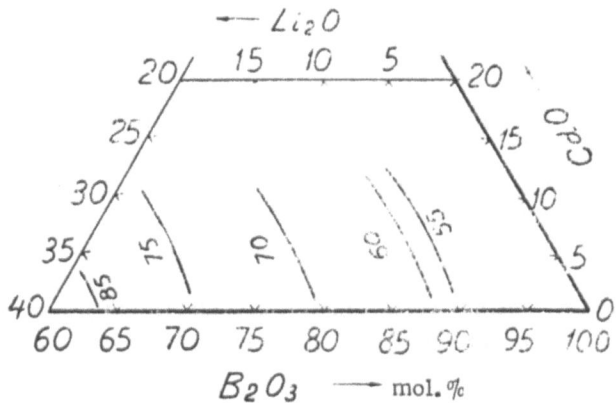

Fig. 107. Thermal expansion ($\alpha \cdot 10^{-7}$) diagram of glasses with the composition $B_2O_3 - Li_2O - CdO$.

The $B_2O_3 - Li_2O - BeO$ System.

The values of the x-ray transparency of the glasses are presented in Table 13 and the compositions of the glasses in Table 9 (appendix).

The absorption of x-rays by plates of the given glass 1.0 cm thick was $100 - X = 22\text{-}26\%$ and for 0.1 cm plates, $0.2\text{-}0.26\%$.

The $B_2O_3 - Li_2O - MgO$ System

The values of the x-ray transparency of the glasses are presented in Table 14 and the compositions of the glasses in Table 13 (appendix).

The $B_2O_3 - Li_2O - BeO - MgO$ System

The values of the x-ray transparency of glasses of this system are presented in Table 15 and the compositions of the glasses in Table 14 (appendix).

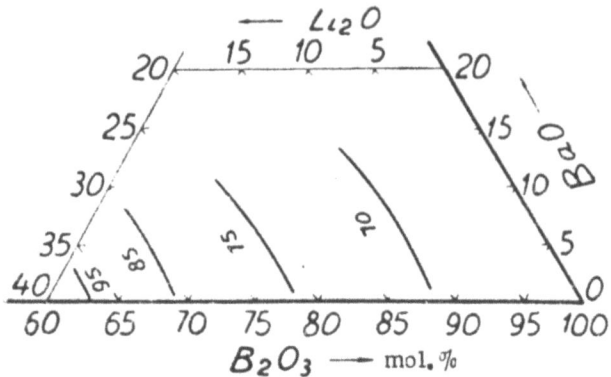

Fig. 108. Thermal expansion $(\alpha \cdot 10^{-7})$ diagram of glasses with the composition $B_2O_3 - Li_2O - BaO$.

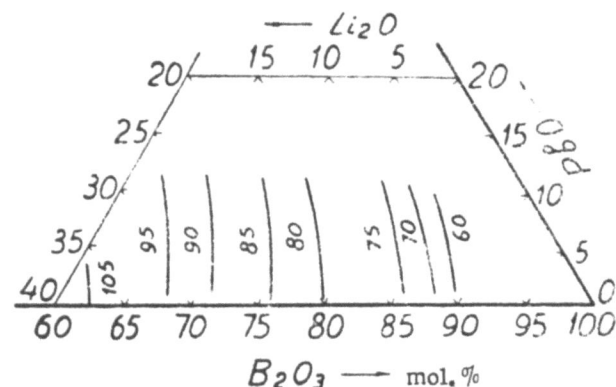

Fig. 109. Thermal expansion $(\alpha \cdot 10^{-7})$ diagram of glasses with the composition $B_2O_3 - Li_2O - PbO$.

Analysis of the calculated data on the x-ray transparency of glasses of the three systems in conjunction with data on their water resistance, thermal expansion coefficients, and other properties shows that industry is in a position to use glass compositions which possess the required thermal expansion coefficients without a protective layer of lacquer on their surface. This also extends considerably their fields of application in other branches of industry, considering the fact that these glasses have a high transparency to ultraviolet rays and other valuable physicochemical properties.

The $B_2O_3 - Li_2O - BeO - MgO$ System with the Addition of SiO_2 and Al_2O_3

To solve the problems of the effect of the addition of the acid oxides SiO_2, Al_2O_3 and ZrO_2 on the structure and properties of borate glasses, we prepared and investigated a special series of glasses with the composition $B_2O_3 - Li_2O - BeO - MgO$ with the different additives SiO_2, Al_2O_3, and ZrO_2 (see the section "The effect of the addition of SiO_2, Al_2O_3, or ZrO_2 on the glass-forming and physicochemical properties of borate glasses)". The values of the x-ray transparency of the series of glasses investigated at a plate thickness of 1.0 cm and $\lambda = 0.1$ A are given in Table 16 and the compositions of the glasses in Table 21 (appendix).

TABLE 12

		BeO	BeO+MgO	CaO	ZnO	SrO	CdO	BaO	PbO
$\alpha \cdot 10^{-7}$ of the glass	from	62,0	53,0	55,0	60,0	65,0	55,0	68,0	60,0
	to	95,0	91,0	106,0	106,0	101,0	85,0	95,0	105,0

TABLE 13

X-ray Transparency of Glasses with the Composition $B_2O_3 - Li_2O - BeO$ at t = 1.0 cm and λ = 0.1 A

Glass No.	X-ray transparency, %	Glass No.	X-ray transparency, %	Glass No.	X-ray transparency, %	Glass type	X-ray transparency, %
13	73,3	28	74,31	38	74,13	№ 46	74,82
14	74,5	29	73,62	39	74,13	Getan - 1	74,13
15	73,45	30	73,62	40	73,29	Getan - 2	73,62
20	74,31	33	75,5	41	73,12	Schleede and Wellmann	75,35
21	73,8	34	75,2	42	75,85	Lindeman	75,17
22	73,62	35	74,82	43	75,35	Botvinkina	74,0
26	74,82	36	74,7	44	75,35	—	—
27	74,5	37	74,82	45	74,82	—	—

TABLE 14

X-ray Transparency of Glasses with the Composition $B_2O_3 - Li_2O - MgO$ at t = 1.0 cm and λ = 0.1 A

Glass No.	X-ray transparency, %	Glass No.	X-ray transparency, %	Glass No.	X-ray transparency, %	Glass No.	X-ray transparency, %	Glass No.	X-ray transparency, %
9	73,84	17	73,84	23	74,17	29	72,16	35	73,17
10	73,67	18	73,60	24	74,08	30	72,0	36	73,05
13	75,29	19	73,35	25	73,92	31	74,91	37	73,32
14	74,43	20	73,22	26	73,72	32	74,10	38	72,05
15	74,18	21	75,18	27	73,36	33	73,96	39	72,00
16	74,12	22	74,47	28	73,16	34	73,71	40	71,84

TABLE 15

X-ray Transparency of Glasses with the Composition $B_2O_3 - Li_2O - BeO -$ $-MgO$ for a Plate Thickness of 1 cm and $\lambda = 1. \lambda$ A

Glass No.	X-ray trans-parency, %	Glass No.	X-ray trans-parency, %	Glass No.	X-ray trans-parency, %	Glass No.	X-ray trans-parency, %	Glass No.	X-ray trans-parency, %
30	74,36	120	74,05	188	75,34	271	76,42	368	74,75
31	77,15	128	75,46	210	73,77	278	75,05	405	73,16
38	75,76	135	74,02	218	75,17	308	74,88	420	73,13
46	77,2	136	77,06	225	73,74	315	74,02	428	74,52
53	75,72	143	75,43	226	76,59	323	74,84	435	73,10
61	77,26	150	74,0	233	75,14	330	73,83	443	74,49
68	75,71	151	76,86	240	73,70	331	76,16	450	73,09
75	74,26	158	75,4	241	76,43	338	74,81	451	75,90
76	77,15	165	73,9	248	75,10	346	76,15	458	74,46
83	75,69	166	76,75	255	73,68	353	74,79	480	72,90
91	77,02	173	75,37	256	74,78	361	76,19	548	74,15
98	75,53	181	76,8	—	—	—	—	—	—

TABLE 16

X-ray Transparency of Glasses with the Composition $B_2O_3 - Li_2O - BeO -$ $- MgO$ with Small Additions of SiO_2 and Al_2O_3 (at t = 1,0 cm and $\lambda =$ $= 0,1$ A)

Glass No.	X-ray trans-parency, %	Glass No.	X-ray trans-parency, %	Glass No.	X-ray trans-parency, %	Glass No.	X-ray trans-parency, %
1	74,81	46	73,09	68	73,55	82	73,62
2	74,25	47	72,59	69	72,05	83	74,18
4	73,96	58	72,80	71	72,53	84	74,25
5	73,44	59	72,30	72	73,85	85	74,36
7	73,40	60	73,95	73	74,15	—	—
44	76,96	67	74,09	74	71,58	—	—

The results show that small additions of the oxides SiO_2 and Al_2O_3 to the glasses produced little fall in the x-ray transparency of the glasses at $\lambda = 0,1$ A.

Series of glasses with the composition $B_2O_3 - Li_2O - BaO$ (PbO) are good absorbers for x-rays. The compositions of the glasses are presented in Tables 19 and 20 (appendix).

ABSORPTION OF THERMAL NEUTRONS BY GLASSES

A series of the glass compositions investigated, especially the compositions $B_2O_3 - Li_2O - CdO$, can also act as absorbers of and shielding against thermal neutrons and γ-rays.

Calculation of the transparency to,or absorption of,slow (thermal) neutrons by the glasses follows a similar rule to that for calculating the transparency to x-rays [82 − 87, 157]:

$$\frac{I}{I_0} = e^{-\mu t} = e^{-\rho t \left(\frac{N\sigma}{\omega}\right)} = e^{-\rho t \omega}.$$

In the case of transmission through glass

$$\frac{I}{I_0} = e^{-\rho t \Sigma (f_i \, \omega_i)} \ .$$

A is the neutron absorption

$$A = 1 - \frac{I}{I_0} \ .$$

Here I_0 is the original intensity of the neutron beam, I is the intensity of the neutron beam after passing through t cm of medium, μ is the absorption coefficient in cm^{-1}, ω is the mass absorption coefficient, ρ is the density of the material (glass) in g/cm^3, N is Avogadro's number = $0.023 \cdot 10^{23}$ atoms and σ is the absolute cross section in cm.

In the case of absorption of slow neutrons, the role of metallic cadmium is analogous to the role of lead for x-rays; therefore, the efficiency of thermal (slow) neutron absorbers is compared with the efficiency of cadmium:

$$\mu_{Cd} t_{Cd} = \mu_{Gl} t_{Gl}$$

At $t_{Cd} = 1$, E of the glass = μ_{Cd}/μ_{Gl}; E of the glass is the glass equivalent of cadmium.

At $t_{glass} = 1$, $E_{Cd} = \mu_{Gl}/\mu_{Cd}$, E_{Cd} is the cadmium equivalent of glass.

Calculations show that for a neutron energy of $1/40$ electron volts(ev), i.e., for thermal or slow neutrons, at $\omega_{Li_2O} = 2.99$, $\omega_{B_2O_3} = 12.6$ and $\omega_{CdO} = 11.0$ and at a cadmium equivalent of $Li_2O = 18.2$ and $B_2O_3 = 4.77$, glasses with the composition $B_2O_3 - Li_2O - CdO$ will practically stop thermal (slow) neutrons and due to their high refractive indices and good chemical stability, they may be used not only as optical and electrovacuum glasses, but also for shielding, especially with 1-2% of cerium oxide added to them.

The results of the investigation show that depending on the nature of the oxide MeO, these borate glasses also have the following physical properties: transparency to x-ray − glasses containing the oxides BeO and MgO; absorption of x-rays − glasses containing the oxides BaO and PbO; absorption of thermal neutrons − glasses containing CdO, etc.

The investigations of Rindol [175] to a certain extent characterize the solarization tendency of these glasses. Solarization is the transfer of electrons between ions of different valence under the action of visible and UV radiation. According to Rindol, lithium glasses

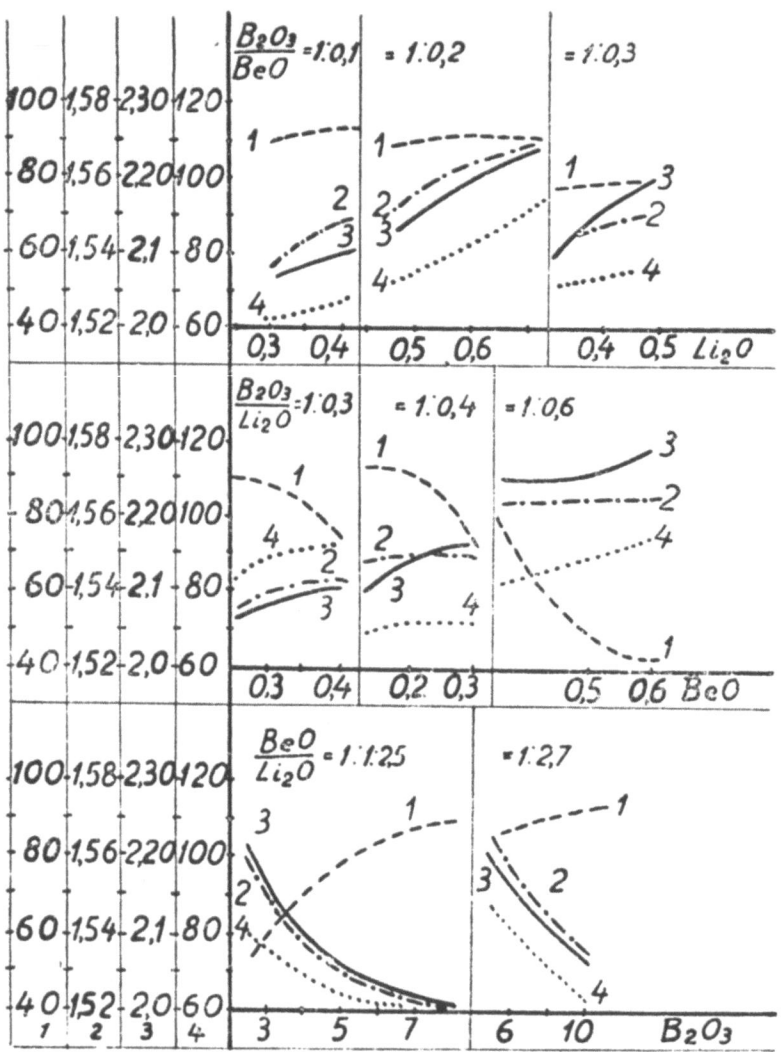

Fig. 110. The effect of separate components of the $B_2O_3 - Li_2O - BeO$ system on the properties of glass: 1) % solubility in water, 2) N_D, 3) D, g/cm³, 4) $\alpha \cdot 10^{-7}$.

containing rare earth oxides are solarized in the following order (according to the intensity of solarization): CaO < BaO > SrO < MgO (they are not solarized in the presence of PbO).

The solarization of glasses in relation to the nature of the alkaline oxides is in the order: Li-glasses < Na-glasses < K-glasses.

These physical properties of the borate glasses we investigated, together with the other technological and physicochemical properties presented previously, open up even wider prospects for their direct (or with the addition of rare earth oxides) use by industry and scientific organizations.

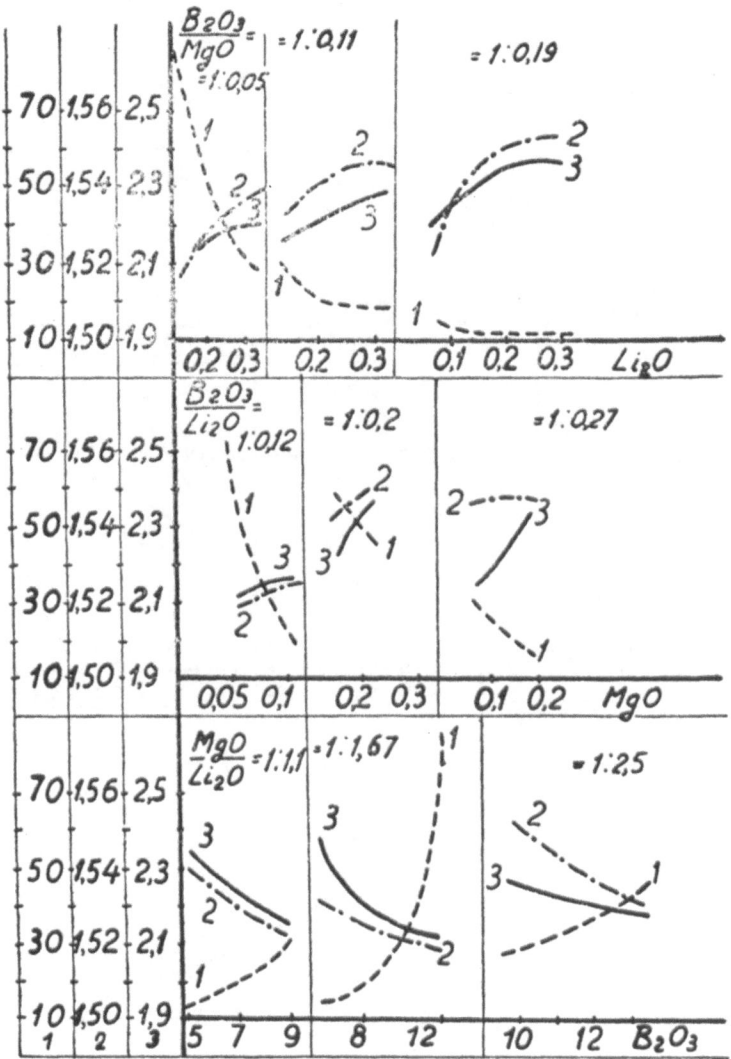

Fig. 111. The effect of separate oxides in the $B_2O_3 - Li_2O - MgO$ system on the properties of glass: 1) % solubility in water, 2) N_D, 3) D, g/cm^3.

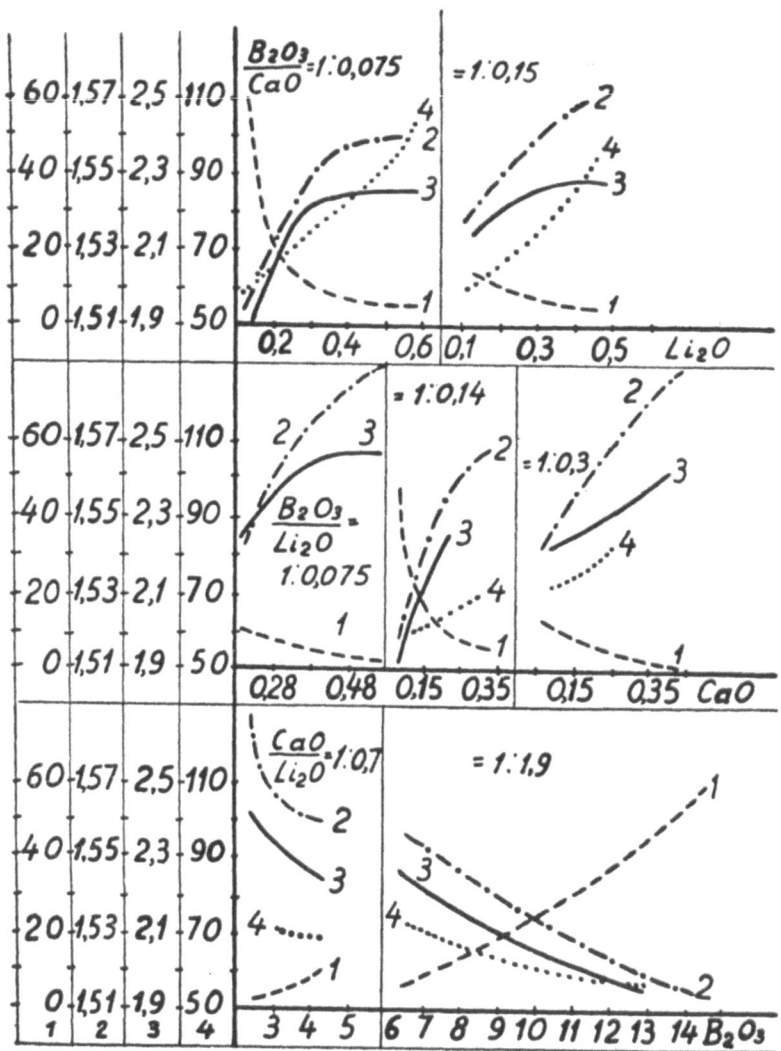

Fig. 112. The effect of separate oxides in the $B_2O_3 - Li_2O - CaO$ system of the properties of glass: 1) % solubility in water, 2) N_D; 3) $D \cdot g/cm^3$, 4) $\alpha \cdot 10^{-7}$.

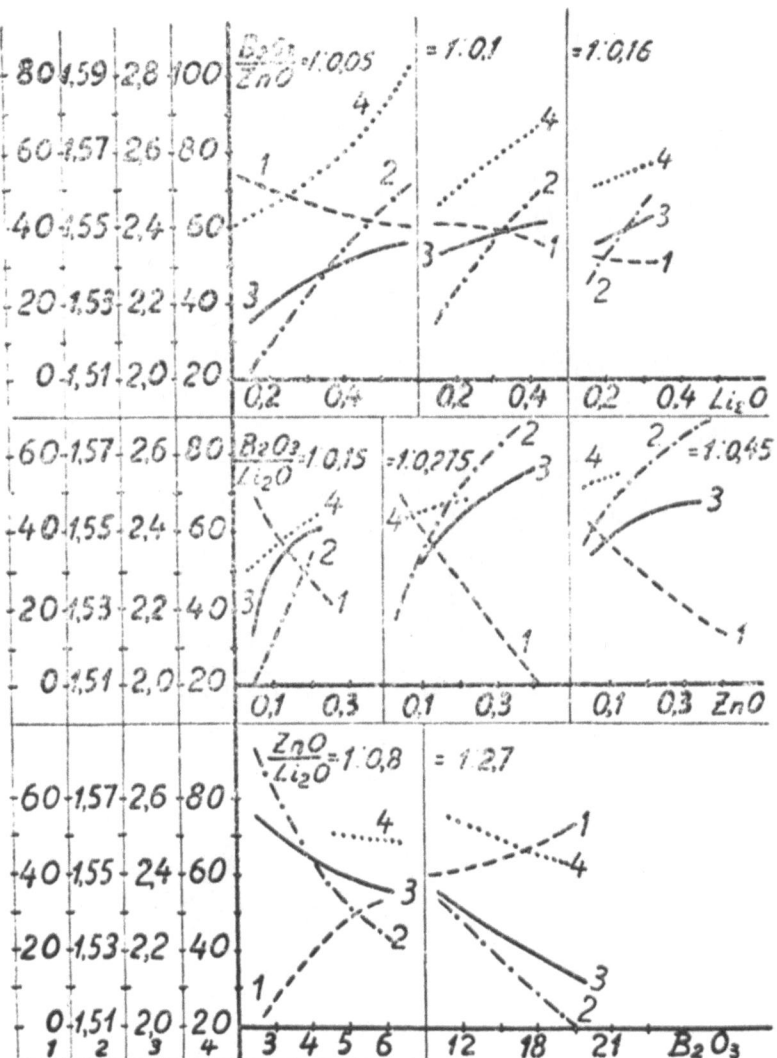

Fig. 113. The effect of separate oxides in the $B_2O_3- Li_2O - ZnO$ system on the properties of glass: 1)% solubility in water, 2) N_D, 3) D, g/cm³, 4) $\alpha \cdot 10^{-7}$.

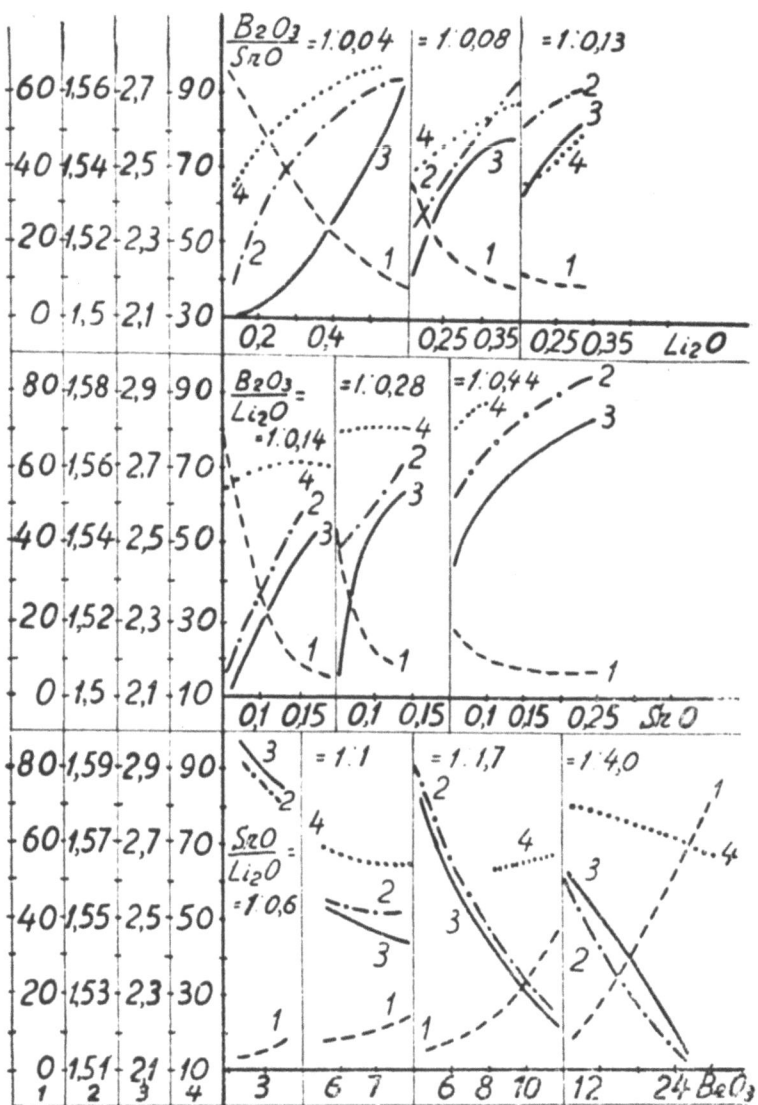

Fig. 114. The effect of separate oxides in the $B_2O_3 - Li_2O - SrO$ system on the properties of glass: 1) % solubility in water, 2) N_D, 3) D, g/cm³, 4) $\alpha \cdot 10^{-7}$.

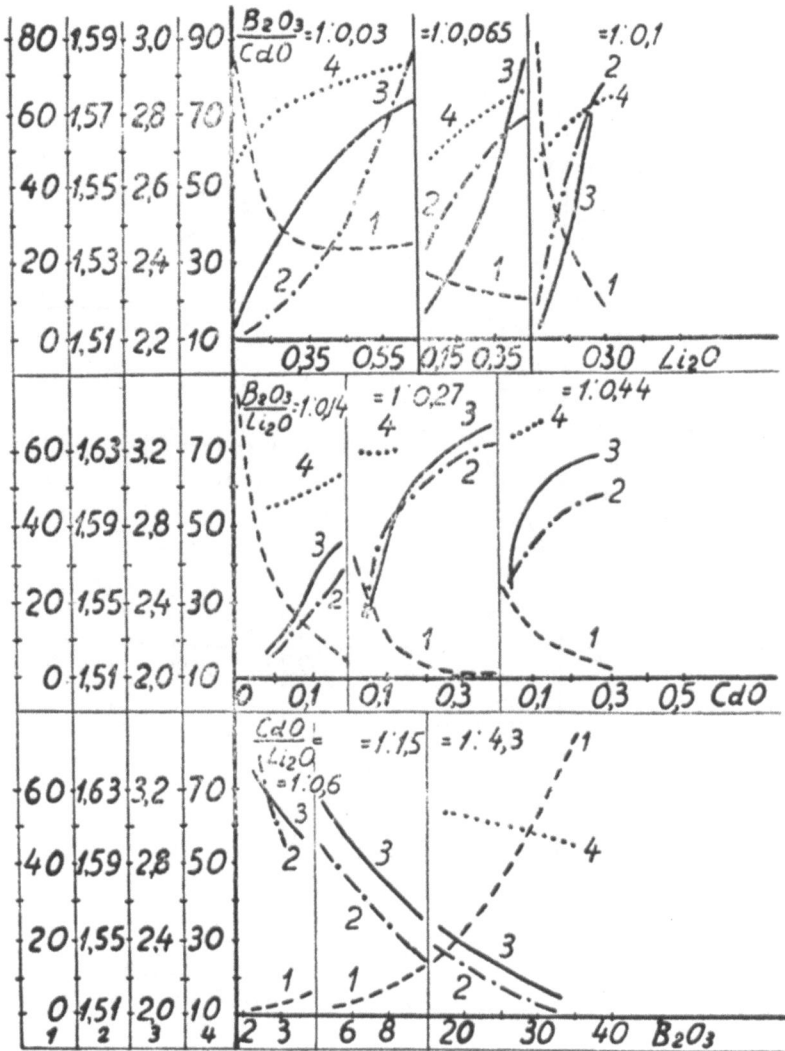

Fig. 115. The effect of separate oxides in the $B_2O_3 - Li_2O - CdO$ system on the properties of glass: 1) % solubility in water, 2) N_D, 3) D, g/cm³, 4) $\alpha \cdot 10^{-7}$.

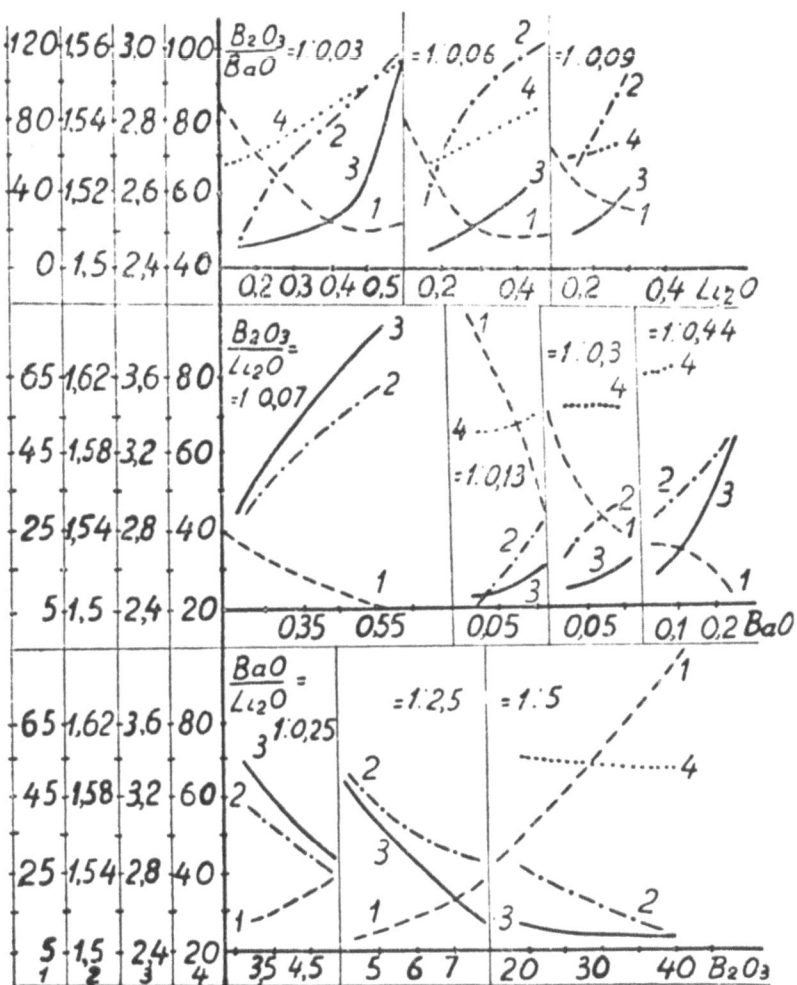

Fig. 116. The effect of separate oxides in the $B_2O_3 - Li_2O - BaO$ system on the properties of glass: 1) % solubility in water, 2) N_D, 3) D, g/cm^3, 4) $\alpha \cdot 10^{-7}$.

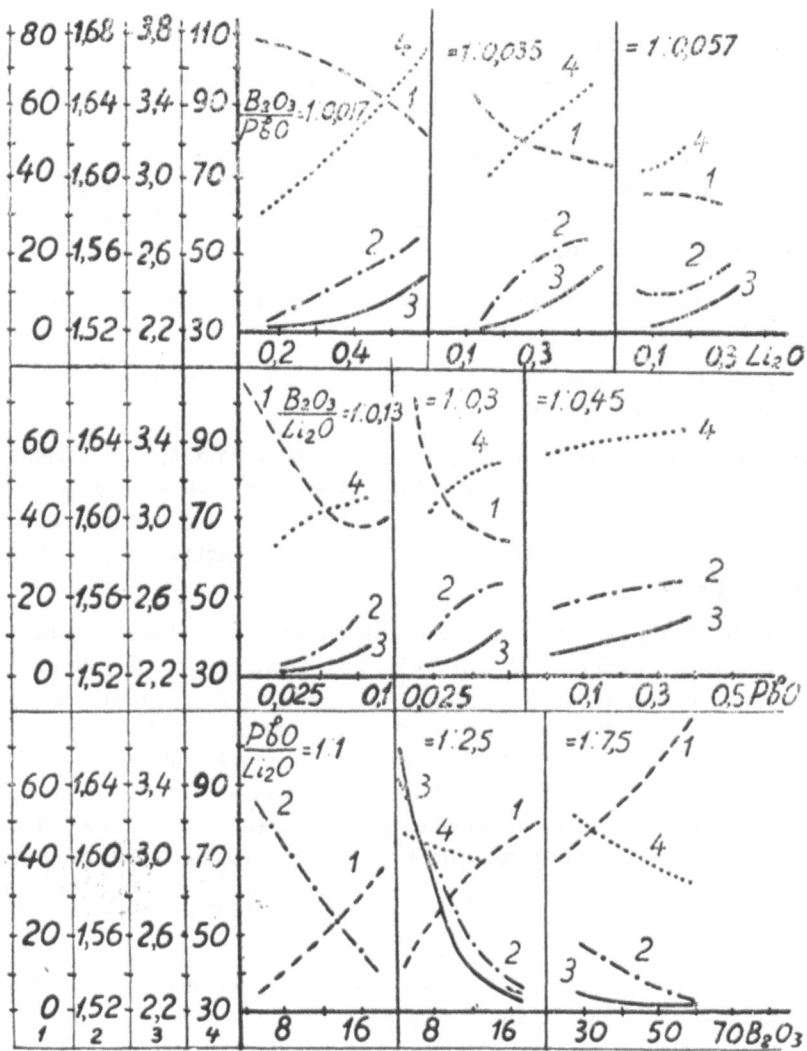

Fig. 117. The effect of separate oxides in the $B_2O_3 - Li_2O - PbO$ system on the properties of glass: 1) % solubility in water, 2) N_D, 3) D, g/cm^3, 4) $\alpha \cdot 10^{-7}$.

THE EFFECT OF THE ADDITION OF SiO$_2$, Al$_2$O$_3$ OR ZrO$_2$ ON THE GLASS FORMING AND PHYSICOCHEMICAL PROPERTIES OF BORATE GLASSES

To solve a series of technological problems for improving the working, physicochemical and mechanical properties of the glasses and also problems of the structure of borate glasses, we also investigated 220 glasses with the composition B$_2$O$_3$ $-$ LiO$_2$ $-$ BeO $-$ MgO with the addition of from 1 to 70 parts by weight of SiO$_2$, Al$_2$O$_3$ or ZrO$_2$.

It is known that at a definite SiO$_2$, B$_2$O$_3$ and Na$_2$O content in the sodium borosilicate glass, the boron and silicon oxides form a so-called submicroheterogeneous structure, which is readily detected by thermal and chemical treatment of the glass. The functions of SiO$_2$, Al$_2$O$_3$ and ZrO$_2$ in glasses with a predominance of B$_2$O$_3$ are unknown.

The extent to which the properties of borate glasses would change if these oxides were added to them is also unknown.

The compositions of the glasses investigated which did not crystallize on cooling are presented in Tables 21 and 22 (appendix).

The maximum temperature for melting the glasses was 1300-1350°. The addition of up to 50 parts by weight of SiO$_2$, Al$_2$O$_3$ or Al$_2$O$_3$ + SiO$_2$ did not essentially change the melting and working conditions of the glasses. The melts were readily drawn, cast and pressed and did not crystallize on cooling. The addition of from 50 to 70 parts by weight of SiO$_2$ or Al$_2$O$_3$ resulted in nonfusion of the glass.

In high-boron glasses, the addition of large amounts of SiO$_2$ or SiO$_2$ + Al$_2$O$_3$ was only possible for glasses with an Li$_2$O content above 5%. Low-alkali glasses, i.e., with an Li$_2$O content in the range 2.5-3%, crystallized on cooling. The addition of any quantity of Al$_2$O$_3$ did not produce crystallization of the glass on cooling, even in the case of a low alkali content. The compositions of the crystallization products in forcibly crystallized glasses were determined, as in the case of glasses without additives, by the quantitative content of the oxides MeO and also by the presence of free B$_2$O$_3$ and the previous thermal treatment of the glass. Thus, the addition of Al$_2$O$_3$ did not have an essential effect on the crystallization of borate glasses.

The addition of SiO$_2$ to glasses containing $\sim$ 90 weight % of B$_2$O$_3$ promoted crystallization and the production of liquation.

The addition of up to 50 parts by weight of ZrO$_2$ did not worsen the quality of the melt and the clarification of the glasses or the working conditions but led to opaque glass. The degree of opacity depended on the concentration of ZrO$_2$ in the glass.

We also investigated the physicochemical properties of the clear glasses obtained with SiO$_2$, Al$_2$O$_3$ or SiO$_2$ + Al$_2$O$_3$ added and compared them with the properties of the starting glasses without additives. The results are presented in Tables 21 and 22 (appendix).

The addition of SiO$_2$ or Al$_2$O$_3$ did not essentially change the properties of the starting glasses for the better and the addition of SiO_2 alone sometimes made them somewhat worse. Evidently, under certain conditions, SiO_2 and also ZrO_2 caused the formation of a submicro-heterogeneous structure in these glasses. The addition of SiO_2 did not always promote the formation of borate glasses with a stronger structure and in some cases disintegration of the basic structure became possible, as shown by a worsening of a number of physicochemical properties of these glasses in comparison with the starting ones.

The role of Al_2O_3 is less clear. It might be supposed that the strong tendency of Al^{3+} to form $[AlO_4]^{4-}$ tetrahedra [9, 10] even in the presence of boric oxide causes the boron to remain tricoordinate and, consequently, to a certain extent helps to maintain the constancy of the glass structure and also its properties.

In addition, the possibility of adding up to 50 parts by weight of SiO_2, Al_2O_3 or ZrO_2 to borate glasses without essentially changing the extremely important physicochemical and physical properties in most cases, considerably extends the field of application of these glasses, lowers their cost, facilitates the solution of a series of problems in the structure of borate glasses, and the character of the change in their properties with change in composition, and also indicates a route for the possible application of natural unprocessed borate raw materials in the production of different glasses and enamels.

Almost analogous conclusions may be drawn about glass formation in the $B_2O_3 - Li_2O - ZnO$ (CdO, SrO) systems with up to 50 parts by weight of SiO_2, Al_2O_3, ZrO_2 and TiO_2 added.

CALCULATION OF THE PROPERTIES OF BORATE GLASSES

A voluminous literature is devoted to the calculation of the properties of silicate and borosilicate glasses. Some start from the rule of transition from quantity to quality and others reduce qualitative differences in the properties of glass to pure quantitative differences and the whole to a simple arithmetic sum of the parts. There is no doubt that some properties of a series of glass compositions are, within certain limits, approximately additive, but this is no basis for a broad theoretical generalization that the whole (glass) is a simple sum of the separate components.

The teaching of D. I. Mendeleev [2] had a profound effect on the mechanical approach to the principle of additivity. Even in 1864 in the monograph "Glass Production" and later in his working notebooks, D. I. Mendeleev wrote that glass is a complex system of indeterminate composition and that "the complex is not a simple mechanical or arithmetical sum of its simpler parts," that "the complex is a qualitatively new formation in comparison with its component parts," and that "it would be a great error to represent a chemical compound as a mechanical combination of atoms in space." As an example of the nonadditivity of certain properties, in particular, the specific volumes of complex substances in relation to the specific volumes of its component parts, D. I. Mendeleev presented the specific volumes of the following minerals:

$CaO \cdot B_2O_3 \cdot 2SiO_2$ (danburite — specific gravity 2.95) has a specific volume of 83.4, while the total of the specific volumes of the component oxides is 101.8; $CaO \cdot 2MgO \cdot 3SiO_2$ (pyroxene — specific gravity 3.467) has a specific volume of 96.7, while the total of the specific volumes of the component oxides is 107.0.

The component parts of the whole, which are combined together, change their properties as they (the oxides) are not simply mechanically added to each other, but interact chemically, affecting each other and even penetrating into each other.

The widely known "rules of additivity," proposed by Winkelmann and Schott in 1894, were based on the premise that the properties of glass are directly additive and are made up of the properties of the component oxides, allowing for their percentage weight content in the glass [116]. Winkelman and Schott proposed the following formula for calculating the additivity of the properties of glass

$$P = \frac{a_1 p_1 + a_2 p_2 + a_3 p_3}{a_1 + a_2 + a_3} = \frac{a_i \cdot p_i}{\Sigma a_i}$$

or

$$P = a_1 p_1 + a_2 p_2 + a_3 p_3 + \ldots$$

Here P is a property of the glass, p is the properties of each oxide, and a is the percent content of each oxide.

For characterizing the properties of each oxide, they proposed specific numerical coefficients referring to 1 wt. % of the oxide. This means that the properties of the glass are determined only by the total of the properties of the weight amounts of the components and do not depend on the nature of the existing oxides, their quantitative ratio to each other, and the changes which they undergo during glass formation.

The properties (according to the formula of Winkelmann and Schott) are expressed graphically on composition — property diagrams as straight lines, but this is by no means confirmed by experiments on thermal expansion, density and certain other properties of glass.

In practice, some investigators frequently identify the properties of separate oxides in the free state with their properties in the glass. Many have determined the coefficient for calculating the properties of oxides in the glass, either by accepting the property in complex glasses equal to the property in quartz glass and determining the properties of the other oxides by difference or by determining the properties of the oxides (coefficients) by a purely arbitrary combination of numbers merely by solving certain mathematical equations with n unknowns. Such an arbitrary "development" of Winkelmann and Schott's additivity rule has resulted in the situation that now for each property, even of glasses of similar composition, there are up to 10-25 series of coefficients and their number tends to grow.

In the 1920's, Gelhoff and Thomas examined the accuracy of the Schott and Winkelmann constants and established that, as a rule, the coefficients were variables and depended on the quantitative amount of the oxides in the glass. For the calculation of glass properties they proposed the "method of differences" or "variable constants" and drew up appropriate tables, making it possible to determine how the value of a given property of a glass changed when each weight percent of SiO_2 in it was replaced by the same amount of another oxide for ranges of the total content of the latter in the glass of from 0-5, 5-10%, etc. They took the properties of the predominating and main glass-forming oxide, SiO_2, as constant and ignored the importance of the change in structure of the glass and the effect of the medium on these changes [95, 161].

In the 1930's, Gilard and Dubrul proposed new coefficients for the calculation of glass properties and the following formula for their calculation [162, 163]:

$$y = \Sigma(ax + bx^2).$$

The curves on the composition — property diagrams were parabolas. The coefficients for a series of oxides were variable and those for SiO_2 remained constant. For this reason and a series of other reasons also, the coefficients of Gilard and Dubrul have an extremely limited application; the thermal expansion coefficients they gave were only for the temperature range 100-130°. Their coefficients were found to be inapplicable to lead and borate glasses, especially those with an alkali content of less than 5 wt. %. Gilard and Dubrul derived their coefficients from a study of only 136 glass compositions and the calculated and experimental data showed considerable discrepancies.

An unusual method of calculating the refractive index of glass was proposed by Sun [164].* He proposed variable coefficients (N_D) of the oxides as functions of the ratio of the number of silicon atoms to the number of oxygen atoms in the glass, independent of the quantitative content of all the components. This method is exceptional.

---

* As in the original Russian. Sun's work is cited in [71, 83, 85, 156].

In 1947 L.I. Demkina published her interpretation of methods of calculating the optical properties of glass. She started from the fact that the optical properties of glass obey the additivity rule to a certain extent, but that it is wrong to identify the weight amounts of substance in the glass with the number of gram-molecules and it is also wrong to ignore the peculiarities of structure of the separate oxides in the glass connected with the geometrical dimensions of the atoms, their surroundings, chemical affinity, and other factors which cannot be considered as yet, but which undoubtedly have an effect on the resultant value of the optical constants [165, 182].

L. I. Demkina proposed the following formula for calculating the properties of glass:

$$P = \frac{\frac{a_1}{S_1} p_1 + \frac{a_2}{S_2} p_2 + \ldots}{\frac{a_1}{S_1} + \frac{a_2}{S_2} + \ldots} = \frac{\sum \frac{a_i}{S_i} p_i}{\sum \frac{a_i}{S_i}} .$$

Here P represents the properties of the oxides, a_i is the number of moles of oxides in the glass, and S_i are structural coefficients depending on structural changes of the oxides in the glass.

A. A. Appen [33-35, 105, 164] investigated methods of calculating the properties of silicate glasses based on definite ideas on the structure of the glass and factors affecting changes in the properties, and derived the following formula:

$$X = \frac{\Sigma(a_0 \cdot \Phi_0)}{\Sigma a_0} = \frac{\Sigma(a\% \cdot \Phi_0)}{100} .$$

Here X is the property of the glass sought, Φ_0 is the value of the coefficient (factor), a_0 is the quantitative amount of the oxide in the glass, expressed in molecular fractions, and $a\%$ is the same, expressed in mol. %.

A. A. Appen gave the variable values of the partial numbers for SiO_2, TiO_2, B_2O_3, CdO and PbO. The calculation method and the partial numbers he proposed were found to be applicable to silicate glasses with a definite SiO_2 content.

Successful developments in glass manufacture and an increase in the number of oxides used in definite weight amounts for the production of glasses for various purposes and for the production of new types of glass articles, operating under very complex conditions, has produced an urgent need for careful study of the properties of glasses and methods of calculating them, in particular, the properties of and calculation methods for purely borate glasses, and for a development of available scientific work on the properties of silicate glasses as applied to borate and other glasses.

It is a very gross error to transfer mechanically the methods of calculating properties, investigated for one system at a definite content and ratio of the separate oxides, to glasses of another system or even of the same system, but with a sharp change in the content of oxides and their ratio to the main glass-forming oxides. So-called "universal" coefficients, suitable for calculating the properties of any glass compositions, do not exist and cannot exist in nature. The present state of science and the existence of suitably staffed and equipped laboratories makes it possible to solve these problems successfully by analyzing the results of investigations from a purely scientific point of view.

The present section is devoted to an analysis of the results of investigating the properties of borate glass and methods of calculating them. The results are based on a considerable amount of experimental material and also the experimental material of other investigators [58, 59, 75-77, 92, 155, 166].

. We studied mainly the properties of B_2O_3 in glasses of the systems investigated. The compositions of the groups of glasses investigated and their properties are presented in tables in the appendix.

On analyzing the results (statistical data and also curves on composition − property diagrams and other material), we arrived at the conclusion that the partial numbers (coefficients) expressing the properties of the oxides in the glass (both glass-forming and "modifying" oxides) are not constants since, depending on their content and their ratio to each other and to B_2O_3, there may exist compounds of the type $aMe_2O \cdot bB_2O_3$, $aMeO \cdot bB_2O_3$ and $aMe_2O \cdot bMeO \cdot cB_2O_3$ and others, where the parameters a, b, and c are variable values.

There is also a certain amount of free B_2O_3 in high-boron glasses. As a rule, the curves of the change in properties are characterized by the presence of extrema (maxima and minima), depending on the given conditions. This means that it is necessary to take into account not only a disordered, but also a partially ordered structure (with a variable degree of ordering) of a continuous steric chain skeleton of the borate glass and also the fact that certain of its component parts, mainly B_2O_3 in this case, can be present in the glass in a different and variable coordination state (difference in the ratio $BO_4 : BO_3$). This is also confirmed by the difference in the partial numbers (coefficients) for B_2O_3 in different systems of glasses, depending on the nature of the other oxides present in the glass and their quantitative content.

To determine the partial number (coefficient) for any oxide and to allow for its effect on the properties of the glass, it is necessary to start from the value of this coefficient for the given oxide in the free state, from the changes that the oxide undergoes in the glass and from its effect on the properties of the glass according to composition − property diagrams. Mathematical equations only help to express this quantitatively and are not an aim in themselves.

The main role in all the structural changes in the borate glasses investigated is played by the predominating glass-forming (and sole acid) oxide, B_2O_3. The possible participation of individual divalent cations in the formation of the lattice of the glass structure (in particular Be^{2+}) is not so significant. The role of the other oxides in the glass is largely expressed as a definite effect on the glass structure by the formation of definite chemical compounds with boron and a change in the $BO_4 : BO_3$ ratio in the glass. Therefore, we consider that it is procedurally correct to take the partial numbers (coefficients) for B_2O_3 as variables depending on the system and the ratio $\dfrac{MeO + Li_2O}{B_2O_3}$ in it, and those for the other oxides provisionally as constants, subject to further study. This is further confirmed by the fact that one encounters such a multiplicity of different values for the coefficients for B_2O_3.

Taking as a basis the partial numbers for Me_2O and MeO, proposed by A. A. Appen as covering a large number of oxides, we obtained partial numbers (coefficients) for B_2O_3 in each of the given glass compositions, including those for glasses studied by other investigators [58, 59, 75-77, 92, 155, 166]. The content of other oxides in the glass and their quantitative ratio to B_2O_3 were considered invariable.

The expression of the values of the partial numbers for B_2O_3 required complex equations of the cubic parabola type or other complex powers and logarithmic functions. The somewhat simplified equation may be represented in the following form:

$$X_{B_2O_3} = K_{B_2O_3} \pm A(100 - B_2O_3),$$

where $X_{B_2O_3}$ is the required quantitative value of the given property for B_2O_3; $K_{B_2O_3}$ is the given property for B_2O_3 as a glass in the free state, and A is a constant for the given type of glass at a definite value of $\dfrac{MeO + Li_2O}{B_2O_3}$.

A expresses the quantitative effect of the separate oxides on the change in properties of B_2O_3 in glass of the given composition. The constant A may be a simple or complex power and logarithmic function, especially in the calculation $\alpha \cdot 10^{-7} B_2O_3$.

The types of equations we took are not arbitrary, but are determined by the actual character of the experimental curves. We expressed the component content in mole fractions.

DENSITY OF GLASS

The compositions of the glasses and also the density values are presented in Tables 9 and 13-21 (appendix) and also in the appropriate phase diagrams. The results of investigations by other authors were also considered in the analysis.

In the literature, one encounters many coefficients for calculation of the effect of each oxide on the glass properties. It is difficult to analyze the reasons for such a variety, as few authors give the compositions of the glasses for which they obtained the recommended coefficients and precisely the compositions for which they are recommended, and then, without grounds, the authors pretended that they were "universal."

TABLE 17

Range of Density Values for the Glasses Investigated (g/cm^3)

Density	BeO	MgO	CaO	ZnO	SrO	CdO	BaO	PbO
D of glass (experimental), g/cm^3 from	2,00	2,05	1,90	2,14	2,10	2,16	2,45	2,12
to	2,28	2,38	2,47	2,80	3,10	3,60	3,91	4,35
$V_{B_2O_3}$, cm^3/mole from	36,0	34,7	38,5	35,5	34,9	34,0	31,4	36,4
to	33,2	30,5	32,8	31,0	26,5	23,7	24,1	27,2

The density of B_2O_3 was calculated from the formulas:

$$D_{B_2O_3} = 1,83 + A(100 - B_2O_3)\ g/cm^3$$
$$V_{B_2O_3} = 38,0 - A(100 - B_2O_3)cm^3/mole$$

The constants A vary for different types of glass and even for the same type of glass at different quantitative oxide contents and $\dfrac{Li_2O + MeO}{B_2O_3}$ ratios.

Data for the calculation of D and V are given in Table 21.

The absolute values of $D_{B_2O_3}$ for all the types of glass were within the range 1.90 to 4.35 g/cm^3 and $V_{B_2O_3}$ from 38.5 to 2.37 cm^3/mole (Table 17).

On comparing the values of the coefficients of different oxides in the free state and in the glass, it is obvious that these values increase considerably for B_2O_3 and Li_2O and change somewhat less for the oxides MeO. This, undoubtedly, is also a reflection of the comparative extent of their effect on the glass structure.

REFRACTIVE INDICES OF GLASS

The compositions of the glasses and also the values of the refractive indices determined are presented in Tables 9 and 13-21 (appendix) and also in the appropriate diagrams.

The absolute values of $N_{B_2O_3}$ in the glasses investigated were from 1.410 to 1.560 (Table 18).

TABLE 18

Range of Refractive Index Values for the Glasses Investigated

N_D		BeO	MgO	CaO	ZnO	SrO	CdO	BaO	PbO
Glass	from	1,520	1,515	1,515	1,508	1,508	1,513	1,503	1,515
	to	1,574	1,577	1,599	1,596	1,596	1,653	1,615	1,715
B_2O_3	from	1,440	1,485	1,475	1,475	1,475	1,480	1,450	1,460
	to	1,520	1,560	1,520	1,520	1,510	1,520	1,485	1,520

The equation for calculation of $N_{DB_2O_3}$ for three- and four-component glass compositions and for the binary composition $B_2O_3 - Na_2O$ will be as follows:

$$X = 1,462 + AX,$$

where $X = 100 - B_2O_3$ for glasses with the oxides BeO, CaO, ZnO, SrO and BaO; A is a variable value and depends on the ratio $\dfrac{MeO + Li_2O}{B_2O_3}$; X = MgO for magnesium glasses; X = CdO or PbO when their content in the glass is more than 25 mol. %. The different meanings of X indicates the difference in the degree of effect of the oxides MeO on the structure and, consequently, on the properties of the glasses.

For a binary glass with the composition $B_2O_3 - Na_2O$, N = 1.458 + ANa_2O; A = 0.001-0.002. The values of $N_{B_2O_3}$ are presented in Table 21.

Analysis of the results shows that the refractive index is also a structure-sensitive property.

According to literature data [166], the values of $\alpha \cdot 10^{-7}$ for B_2O_3 (glass) approximately equal 152. With an increase in temperature, B_2O_3 first contracts somewhat and then expands sharply. The value of $\alpha \cdot 10^{-7}$ of B_2O_3 varies insignificantly up to a temperature of 175° and then changes sharply within the following limits:

$$\alpha_{B_2O_3} = 152 + \Delta t^n;$$

n = 1.16 for temperatures from 175 to 200°, and n = 1.5 for temperatures from 100 to 250°.

Two-component glasses with the composition $B_2O_3 - Li_2O$ [166] have a clearly expressed minimum in $\alpha \cdot 10^{-7}$ at an Li_2O content in the glass of up to 20 mol. % or up to the ratio O/B $\cong$ 1.63. At Li_2O content of more than 20 mol. % and O/B > 1.63, the value of α for the glass increases.

For two-component glasses with the composition $B_2O_3 - Na_2O$, $\alpha \cdot 10^{-7}$ of the glass has a sharply expressed minimum at an Na_2O content of ~ 17 mol. % and O/B $\cong$ 1.63. With an increase in the Na_2O content, $\alpha \cdot 10^{-7}$ of the glass increases to a definite maximum at a ratio $\frac{Na_2O}{B_2O_3}$ = 0.5 and O/B = 1.98 and then falls again.

The minimum corresponds to a glass with the composition $Na_2O \cdot 5B_2O_3$ and the maximum to $Na_2O \cdot 2B_2O_3$. Similarly to Li_2O, glasses with the composition $B_2O_3 - K_2O$ have this minimum at a K_2O content of ~ 20 mol. %. With an increase in the K_2O content, $\alpha \cdot 10^{-7}$ of the glass increases. This indicates that with the introduction of 17-20 mol. % alkalies into the glass, the glass does not expand with a rise in temperature, but contracts in the first stage; then, with a higher content of the oxides Me_2O and MeO in the glass, the heat introduced produces a corresponding expansion of the glass.

The presence of an expansion minimum at a glass composition of $Na_2O \cdot 5B_2O_3$ indicates that the structural groups XY_4 (X = BO_4, Y = BO_3) are formed at this ratio. However, this contradicts the statement [92] that boron is predominantly in the BO_4 form at this glass composition.

The minima mentioned above were also found when the oxides MeO were introduced into the glass (and not just in the presence of Me_2O). For example, a similar minimum in $\alpha \cdot 10^{-7}$ of the glass was also found for the composition $B_2O_3 - PbO$ at a PbO content of 30 mol. %. With an even greater increase in the PbO content, $\alpha \cdot 10^{-7}$ of the glass also increased steeply.

This makes it possible to derive a general rule on the behavior of binary borate glasses during the determination of their expansion coefficients. With an Me_2O or MeO content in the glass within definite limits, as has already been stated, in the first stage there is contraction of the glass structure with a rise in temperature and, evidently, a decrease in the interatomic distance; then, with an increase in the oxide content and also with a rise in temperature, there is expansion of the glass, loosening of the structure and, evidently, an increase in the interatomic distance. All this is an indirect criterion for establishing the direction of the change in the quantitative ratio BO_4/BO_3 in the glass.

The character of the expansion curves for the ternary systems investigated was approximately the same. The expansion curves were relatively flat at first and then became steeper with a further increase in the content of the oxides Me_2O and MeO.

Expressing the value of $\alpha_{B_2O_3}$ by a definite equation is an extremely complex problem, since the value changes over very wide limits: from $+152$ to -167 [166]. In this case it is necessary to consider most fully such factors as the B_2O_3 content, the nature and quantity of the other oxides, their quantitative ratio to boric oxide, etc.

The value of $\alpha_{B_2O_3}$ can be expressed most accurately by the two very complicated, following equations:

I. $Y + C = -A(X-B)^3$ (an equation of the cubic parabolic type)m where $Y = \alpha \cdot 10^{-7}$; $X = (100 - B_2O_3)$; A, B and C are constants of which C is arbitrary and A and B can be determined by the method of averages.

II. $Y = A \cdot e\, BX - C$ (a complex power function), where $Y = \alpha \cdot 10^{-7}$, C = constant = 152, A and B can be determined.

Such methods of determining $\alpha_{B_2O_3}$ are extremely complex and are suitable only for constructing nomograms, which, in our opinion, should be carried out in collaboration with mathematicians.

We made an attempt to simplify somewhat the method of calculating $\alpha_{B_2O_3}$ and proposed the following formula, which made it possible to determine the value of $\alpha_{B_2O_3}$ with sufficient accuracy:

$$\alpha_{B_2O_3} = \frac{152}{100}\ X - C,$$

where 152 is the value of $\alpha_{B_2O_3}$ for the glass in the free state, X is the % B_2O_3 content of the glass, and C is a constant whose value varies from 85 to 110, depending on the quantitative content of the other oxides and on the nature of the oxide MeO.

The values of the coefficients for calculating $\alpha_{B_2O_3}$ for different glass compositions are presented in Table 21.

It should be noted that the effect of B_2O_3 on the properties of the glass is so great in comparison with the effect of the other oxides that for a series of glasses it is possible to determine the value of α of the glass with sufficient accuracy without knowing the values of α for the other oxides by using the formula

$$\alpha_{gl} = \frac{152}{100}\ B_2O_3 - A(B_2O_3 - 46),$$

where A is a constant, depending on the molecular percent content of B_2O_3 and is a correction for the structural change of B_2O_3 in the glass.

The absolute values of $\alpha_{B_2O_3}$ for the ternary systems investigated were within the range from $+73$ to -65.0 (Table 19) and for the two-component compositions, from $+152$ to -167 [166].

The compositions of the glasses are presented in Tables 9, 13-20, and 22 (appendix) and the values of α for the glasses in the same tables and the corresponding phase diagrams.

α of the glass increased continuously with an increase in temperature. For most of the glasses investigated, the expansion slowed in the temperature range of 320-420°, i.e., close to their softening points, and this is connected with structural changes in the glass (Table 20).

TABLE 19

Range of Expansion-Coefficient Values for Different Glass Compositions

$\alpha \cdot 10^{-7}$		BeO	CaO	ZnO	SrO	CdO	BaO	PbO
Glass	from	62	55	60	65	55	68	60
	to	95	106	106	101	85	95	105
B_2O_3	from	−65	−2,1	+1,0	−6,7	−30	−16,7	+2,1
	to	+53	+26	+45	+36,0	+26,4	+39,0	+11,5

Data for calculating the x-ray transparency of glasses are presented above.

The study of the expansion coefficients of glasses with an inhomogeneous medium, opaque glasses of the enamel and glazing type, is a separate problem since applying the same coefficients to their calculation as for clear glasses is incorrect in principle, yet all the textbooks recommend such a method of calculation for these glasses without any grounds.

We present an example of the mathematical calculation of some typical expansion curves of glasses with the help of the following type of equation:

$$Y = Ae^{-BX} - 152 \quad \text{or} \quad \lg(Y - 152) = A - BX;$$
$$Y = \alpha 10^{-7}{}_{B_2O_3},$$

where X is the amount of B_2O_3 in mol. %, A and B are constants which can be calculated, and 152 is the value of $\alpha \cdot 10^{-7}{}_{B_2O_3}$ in the free state.

Glasses with the Composition $B_2O_3 - Li_2O$ (according to Frenet, 1897 [166])

Sample No.	X	Y	Y+152	$\lg(Y - 152)$
1	9,7	74,0	226,0	2,3541
2	16,7	27,5	179,5	2,2541
Σ	26,4	101,5	405,5	4,6082
3	20,0	10,2	162,2	2,2100
4	25,0	4,7	156,7	2.1951
Σ	45,0	14,9	318,9	4,4051

Let us set up the equation

1. $4,6082 = 2 A + B 26,4 \quad | + |$
2. $4,4051 = 2 A + B 45,0 \quad | - |$

———————————————

$0,2031 = \quad - B 18,6$

TABLE 20

Character of the Change in $\alpha \cdot 10^{-7}$ in the $B_2O_3-Li_2O-BeO$ System with a Change in the Heating Temperature

Glass No.	$\alpha \cdot 10^{-7}$ in the temperature ranges							
	20—120°	12·—220°	220—320°	320—420°	20—120°	20—220°	20—320°	20—420°
13	+ 7,4	+14,0	+21,8	+41,0	+ 7,4	+10,3	+14,4	+21,3
15	+ 0,72	+13,0	+29,6	+39,5	+ 0,72	+10,0	+15,7	+22,3
20	− 8,5	− 1,6	+12,1	+24,0	− 8,5	− 3,3	+ 1,15	+ 6,53
21	−17,6	− 5,4	+ 5,5	+12,0	−17,6	−10,8	+ 5,1	+ 1,37
22	−11,0	− 1,8	+16,0	+25,5	−11,0	+ 6,3	+ 2,85	+ 8,0
26	+ 2,35	+17,7	+34,6	+47,5	+ 2,35	+11,6	+19,3	+25,0
27	+ 7,9	+ 0,98	+ 9,0	+15,7	+ 7,9	+ 1,9	+ 1,15	+ 5,55
28	−13,7	− 3,7	+ 6,1	+17,5	−13,7	− 8,6	+ 4,0	+ 3,8
29	−28,0	−19,0	− 5,5	− 4,8	−28,0	−24,0	−16,9	−11,2
30	−26,6	−13,3	− 0,01	+20,8	−26,6	−20,0	−13,0	+ 1,2
33	+15,1	+20,8	+25,7	+31,5	+15,1	+18,0	+20,5	+23,4
34	+ 5,0	+11,0	+20,0	+26,0	+ 5,0	+ 8,0	+12,1	+15,5
35	−10,2	+ 1,1	+15,8	+25,0	−10,2	+ 2,5	+ 3,15	+ 9,9
36	−30,0	−12,4	+ 0,92	+13,3	−30,0	−20,8	−13,6	+ 7,0
37	−39,0	−26,0	−16,3	− 5,3	−39,0	−33,0	−27,4	−22,0
38	−91,0	−75,0	−58,0	−43,5	−91,0	−83,0	−75,0	−65,3
39	−26,6	−14,6	− 7,3	+ 0,2	−26,6	−20,6	−16,5	−12,1
40	−38,3	−24,4	−12,3	+ 5,3	−38,3	−30,8	−24,8	−16,4
41	+34,5	+41,7	+60,0	+75,0	+34,5	+38,0	+45,7	+53,0
43	+ 2,5	+ 7,7	+11,5	+16,6	+ 2,5	+ 5,2	+ 7,3	+ 9,7
44	−22,0	−10,2	− 2,5	+ 4,5	−22,0	−16,4	−12,0	+ 6,3
45	−26,0	−21,2	− 7,5	+13,5	−26,0	−22,1	−16,3	+ 6,3
46	−29,0	−16,0	− 1,1	+15,0	−29,0	−22,0	−15,3	+ 6,0

Hence
$$B = -0,0109 \text{ and } A = 2,448.$$
$$\lg (Y + 152) = 2,448 - 0,0109 \, X;$$
$$Y = 281.10^{-0,0109 \, x} - 152$$

or
$$Y = 281e^{-0,0251 \, x} - 152.$$

Glasses with the Composition $Na_2O - B_2O_3$ (according to Gooding and Turner 1934 [166])

Sample No.	X	Y	(Y + 152)	lg (Y + 152)
1	2,5	134,0	286,0	2,4564
2	4,0	128,0	280,0	2,4472
3	5,6	111,0	263,0	2,4200
4	8,7	83,0	235,0	2,3711
5	11,1	63,0	215,0	2,3324
6	12,7	49,5	201,5	2,3043
Σ	44,6	568,5	1480,5	14,3314
7	17,4	20,0	172,0	2,2355
8	20,4	11,3	163,0	2,2122
9	22,0	3,9	155,9	2,1929
10	24,0	— 4,2	147,8	2,1697
11	27,0	13,7	138,3	2,1408
12	32,6	—40,0	112,0	2,0492
Σ	143,4	—22,7	889,0	13,0003

Let us set up the equation

$$14,3314 = 6A + B \, 44,6 \quad \left| \; + \; \right|$$
$$13,0003 = 6A + B 143,4 \quad \left| \; - \; \right|$$

$$1,3311 = \qquad - B \, 98,8$$

Hence
$$B = -0,0135 \text{ and } A = 2,489.$$
$$\lg (Y + 152) = 2,489 - 0,0135 \, x;$$
$$Y = 308 \cdot 10^{-0,0135 \, x} - 152$$

or

$$Y = 308e^{-0,0311 \, x} - 152.$$

Glasses with the Composition $B_2O_3 - Li_2O - BeO$ (according to L. Ya. Maze-lev, 1940 [61])

Sample No.	X	Y	Y + 152	lg (Y + 152)
1	29,0	23,4	175,4	2,2440
2	31,5	15,5	167,5	2,2240
3	36,5	9,9	161,9	2,2092
Σ	97,0	48,8	504,8	6,6772
4	41,0	5,5	157,5	2,1973
5	43,5	3,8	155,8	2,1926
6	47,5	−11,2	140,8	2,1486
7	51,0	−16,4	135,0	2,1322
Σ	183,0	−18,3	589,7	8,6707

Let us set up the equation

$$6,6772 = 3A + B97 \quad | -4 |$$
$$8,6707 = 4A + B183 \quad | +3 |$$

$$-0,6967 = \qquad B161$$

Hence $\qquad B = -0,00433 \text{ and } A = 2,366$

$$\lg(Y+152) = 2,366 - 0,00433X;$$

$$Y = 232 \cdot 10^{-0,00433\,x} - 152$$

or

$$Y = 232 \cdot e^{-0,00997\,x} - 152.$$

TABLE 21

Formulas for Calculating Partial Properties of B_2O_3 in Glasses

Systems			Density $V_{B_2O_3}$ = cm³/mole				
B_2O_3	Me_2O	MeO					
B_2O_3	Li_2O	—	$V=38-0,21(100-B_2O_3)$	(59)			
B_2O_3	Na_2O	—	$V=38-0,25(100-B_2O_3)$	(92)			
B_2O_3	Li_2O	BeO	$V=38-0,1\ (100-B_2O_3)$	(61)			
B_2O_3	Na_2O	BeO	$V=38-K\ (100-B_2O_3)$	(70)			
			up to	0,2	0,4	0,65	1,0
			K =	0,24	0,16	0,12	0,09
B_2O_3	K_2O	BeO	$V=38-K(100-B_2O_3)$ (77)	ψ	< 0,65	0,65—1,3	
				K =	0,115	0,005	

TABLE 21 (continued)

Systems			Density $V_{B_2O_3}$ = cm^3/mole					
B_2O_3	Me_2O	MeO						
B_2O_3	Li_2O	MgO	$V=38-0,2$ $(100-B_2O_3)$			(62)		
B_2O_3	Na_2O	MgO	$V=38-0,22$ $(100-B_2O_3)$			(77)		
B_2O_3	K_2O	MgO	$V=38-0,115(100-B_2O_3)$			(77)		
B_2O_3	Li_2O	BeO+ +MgO	$V=38-K(100-B_2O_3)$ (68)	ψ up to $K=$	0,6 0,135	from 0,6 up to 1,0 0,07	from 1,0 up to 1,5 0,55	
B_2O_3	—	BeO+ +MgO	$V=38-0,08(100-B_2O_3)$			(68)		
B_2O_3	Li_2O	CaO	$V=38-K(100-B_2O_3)$ (68)	ψ up to 0,25 from 0,25 to 0,5 $>0,5$	K 0,24 0,16 0,13			
B_2O_3	Li_2O	ZnO	$V=38-K(100-B_2O_3)$ (68)	ψ up to 0,35 $>0,35$	K 0,22 0,14			
B_2O_3	Li_2O	SrO	$V=38-0,27(100-B_2O_3)$		(68)			
B_2O_3	Li_2O	CdO	$V=38-K(100-B_2O_3)$ (68)	ψ up to 0,2 from 0,2 to 0,6 $>0,6$	K 0,27 0,36 0,2			
B_2O_3	Li_2O	BaO	$V=38-K(100-B_2O_3)$ (68)	ψ up to 0,2 from 0,2 to 0,5 $>0,5$	K 0,6 0,36 0,2			
B_2O_3	Li_2O	PbO	$V=38-0,2(100-B_2O_3)$		(68)			

$$\psi = \frac{mMe_2O+nMeO}{B_2O_3}$$

II

Systems			Refractive indices $N_{B_2O_3}$
B_2O_3	Me_2O	MeO	
B_2O_3	Li_2O	—	$N=1,4625+0,0025$ $(100-B_2O_3)$ (59)
B_2O_3	Na_2O	—	$N=1,458 +0,002$ $(100-B_2O_3)$ (92)
B_2O_3	Li_2O	BeO	$N=1,4625-0,0006$ $(100-B_2O_3)$ (61)

Systems			Refractive indices $N_{B_2O_3}$
B_2O_3	Me_2O	MeO	
B_2O_3	Na_2O	BeO	$N = 1{,}480 - 0{,}0025 [100-(B_2O_3+Na_2O)]$ (70)
B_2O_3	K_2O	BeO	$N = 1{,}4625 - 0{,}0006 (100-B_2O_3)$ (77)
B_2O_3	Li_2O	MgO	$N = 1{,}4625 + 0{,}003 [100-(B_2O_3+Li_2O)]$ (62)
B_2O_3	Na_2O	MgO	$N = 1{,}462 + 0{,}003 [100-(B_2O_3+Na_2O)]$ (77)
B_2O_3	K_2O	MgO	$N = 1{,}462 + 0{,}002 [100-(B_2O_3+K_2O)]$ (77) $\psi < 0{,}5$
			$N = 1{,}462 - 0{,}002 [100-(B_2O_3+K_2O)]$ $\psi > 0{,}5$
B_2O_3	Li_2O	$BeO+$ $+MgO$	$N = 1{,}4625 + 0{,}00065 (100-B_2O_3)$ (68)
B_2O_3	Li_2O	CaO	$N = 1{,}4625 + A(100-B_2O_3);$ $CaO < 25\%$, $A = 0{,}00065$ (68) $CaO > 25\%$, $A = 0{,}0013$
B_2O_3	Li_2O	ZnO	$N = 1{,}4625 + A(100-B_2O_3);$ $ZnO < 25\%$, $A = 0{,}00065$ (68) $ZnO > 25\%$, $A = 0{,}0013$
B_2O_3	Li_2O	SrO	$N = 1{,}4625 + A(100-B_2O_3);$ $SrO < 25\%$, $A = 0{,}00065$ (68) $SrO > 25\%$, $A = 0{,}0013$
B_2O_3	Li_2O	CdO	$N = 1{,}4625 + 0{,}00065 (100-B_2O_3);$ $Li_2O > 25\%$ $N = 1{,}4625 + 0{,}003 [100-(B_2O_3+Li_2O)];$ $CdO > 25\%$ $N = 1{,}4625 + 0{,}0013 (100-B_2O_3)$, for other compositions (68)
B_2O_3	Li_2O	BaO	$N = 1{,}4625 + 0{,}0003 (100-B_2O_3);$ BaO и $Li_2O > 35\%$ $N = 1{,}4625 + 0{,}00065 (100-B_2O_3);$ for other compositions (68)
B_2O_3	Li_2O	PbO	$N = 1{,}4625 + 0{,}0003 (100-B_2O_3);$ $Li_2O > 30\%$ $N = 1{,}4625 + 0{,}002 ([100-(B_2O_3+Li_2O)]);$ $PbO > 25\%$ $N = 1{,}4625 + 0{,}00065 (100-B_2O_3)$, for other compositions (68)

III

Systems			Thermal expansion, $\alpha \cdot 10^{-7}{}_{B_2O_3}$ (or $\alpha \cdot 10^{-7}$ of the glass)
B_2O_3	Me_2O	MeO	
B_2O_3	Li_2O	—	$\alpha_{B_2O_3} = 1{,}52 - K(100-B_2O_3);$ $K = 7{,}5$ at $Li_2O < 20\%$ (166) $K = 6$, at $Li_2O > 20\%$

113

Systems			Thermal expansion, $\alpha \cdot 10^{-7}$ B_2O_3 (or $\alpha \cdot 10^{-7}$ of the glass)
B_2O_3	Me_2O	MeO	
B_2O_3	Li_2O	BeO	$\alpha_{gl} = 1{,}52B_2O_3 - K\ (B_2O_3 - 46)$ (61) $\hspace{1cm}$ (20—420°)
B_2O_3	Li_2O	BeO+ +MgO	$\alpha_{gl} = 1{,}52B_2O_3 - K\ (B_2O_3 - 46)$ (68) $\hspace{1cm}$ (20—320°)
B_2O_3	Li_2O	CaO	$\alpha_{B_2O_3} = 1{,}52B_2O_3 - 85;\ CaO > 10$, $Li_2O > 35\%;\ \psi > 0{,}5$ $\alpha_{B_2O_3} = 1{,}52B_2O_3 - 100;\ CaO < 10$, $Li_2O < 35\%;\ \psi < 0{,}5$ (68) $\hspace{1cm}$ (20—320°)
B_2O_3	Li_2O	ZnO	$\alpha_{B_2O_3} = 1{,}52B_2O_3 - 85;\ ZnO > 7{,}5$, $Li_2O > 35\%$;(20—320°) $\alpha_{B_2O_3} = 1{,}52B_2O_3 - 95;\ ZnO < 7{,}5$, $Li_2O < 35\%$ (68)
B_2O_3	Li_2O	SrO	$\alpha_{B_2O_3} = 1{,}52B_2O_3 - 95$ $\hspace{1cm}$ (20—320°) (68)
B_2O_3	Li_2O	CdO	$\alpha_{B_2O_3} = 1{,}52B_2O_3 - 100;\ CdO < 10$, $Li_2O < 20\%;$ $\hspace{0.5cm}$ (20—320°) $\alpha_{B_2O_3} = 1{,}52B_2O_3 - 105;\ Li_2O > 20\%$ $\alpha_{B_2O_3} = 1{,}52B_2O_3 - 115;\ Li_2O > 30\%$
B_2O_3	Li_2O	BaO	$\alpha_{B_2O_3} = 1{,}52B_2O_3 - 95;\ Li_2O < 20\%$ $\hspace{1cm}$ (20—320°) $\alpha_{B_2O_3} = 1{,}52B_2O_3 - 105;\ 30\% > Li_2O > 20\%$ $\alpha_{B_2O_3} = 1{,}52B_2O_3 - 110;\ Li_2O > 30\%$ (68)
B_2O_3	Li_2O	PbO	$\alpha_{B_2O_3} = 1{,}52B_2O_3 - 85;\ Li_2O$ from 20 to 30 mol. % $\alpha_{B_2O_3} = 1{,}52B_2O_3 - 90;\ 30\% < Li_2O < 20\%$ (68) $\hspace{1cm}$ (20—320°)
B_2O_3	—	—	$\alpha_{B_2O_3} = 1{,}52 + \Delta t^n;\ n = 1{,}16$ for $t°$ from 175—200° $\hspace{2cm} n = 1{,}5$ for $t°$ from 200—250° (166)

SUMMARY

1. A short review of modern ideas on the structure of glass has been compiled and the characteristics of the main components of the systems investigated are also presented.

2. More than 1500 melts of mixtures were carried out and the corresponding glass formation diagrams set up for the following systems:

$$1.\ B_2O_3-Li_2O-BeO; \qquad 5.\ B_2O_3-Li_2O-SrO;$$
$$2.\ B_2O_3-Li_2O-MgO; \qquad 6.\ B_2O_3-Li_2O-CdO;$$
$$3.\ B_2O_3-Li_2O-CaO; \qquad 7.\ B_2O_3-Li_2O-BaO;$$
$$4.\ B_2O_3-Li_2O-ZnO; \qquad 8.\ B_2O_3-Li_2O-PbO;$$
$$9.\ B_2O_3-BeO-MgO;$$
$$10.\ B_2O_3-Li_2O-BeO-MgO;$$
$$11.\ B_2O_3-Li_2O-BeO-MgO+SiO_2(Al_2O_3,ZrO_2);$$
$$12-14.\ B_2O_3-Li_2O-ZnO(CdO,SrO)+SiO_2, Al_2O_3, ZrO_2, TiO_2.$$

The glass formation and crystallization sections were determined for all the systems investigated, technological and temperature conditions were developed for melting, annealing, and working, and the degree of volatility of the components during melting was established.

The temperatures at which the glasses began to soften on heating were established and, in accordance with them and also the character of the change in the expansion curves of the glasses, the upper and lower annealing temperature limits were established.

3. A procedure was developed for chemical analysis of boron — lithium — beryllium (magnesium) glasses.

4. We investigated thermographically H_3BO_3 and the carbonates from the oxides Li_2O, BeO, MgO, CaO, ZnO, SrO, CdO, BaO, PbO and their binary and ternary mixtures on an apparatus of the Kurnakov type and obtained corresponding thermograms. Curves for the weight losses during dynamic heating were obtained for the same components and mixtures of them.

5. On the basis of the following series of investigations:

a) the melting points of the systems and determinations of the glass formation and crystallization regions;

b) studies of the technological properties, melting and annealing conditions and the volatility of the components and the results of chemical analysis of the glasses by the procedures we developed;

c) studies of the thermochemical processes and reactions during the heating of mixtures and glass formation;

d) studies of the crystallization of glasses and the composition and characteristics of the crystallization products;

e) investigations of the physicochemical properties of glasses: density, refractive index, chemical stability, thermal expansion, etc., and also a study of literature data, the following were accomplished:

A) a scheme was drawn up for the thermochemical processes and reactions in mixtures of components during heating;

B) the dissociation points of the separate components and their mixtures and the melting points of B_2O_3, $LiCO_3$, and the other carbonates were established;

C) the formation temperatures of eutectic mixtures and borates of various compositions were determined and literature data on the formation of binary eutectics and compounds of B_2O_3 with certain oxides of the MeO type were also determined more precisely;

D) the limiting temperatures for complete loss of volatile components by separate components and their mixtures on heating were established;

E) the main endo- and exothermal effects during the heating of separate components and their mixtures were defined more precisely;

F) the temperature ranges for the completion of glass-formation processes in different systems were established.

6. The formation of eutectics and borates during the heating of mixtures proceeds at lower temperatures than in silicate glasses and, likewise, the formation of a liquid phase (melting point of B_2O_3 ~ 525°, beginning of melting of Li_2CO_3 ~ 650°).

In most of the glasses investigated, the glass formation processes were complete at up to 1200°. At these temperatures there occurred lightening and homogenization of the glass mixture, which had a very low viscosity and a low surface tension. When the glass mixture was cooled, its viscosity rose very rapidly, especially in high-boron compositions. In most cases, the boron — lithium — beryllium and magnesium glasses investigated had a small range of working temperatures, i.e., they were "short." Glasses of the other systems had a greater range of working temperatures. For a series of glasses it was found that the softening point coincided with the temperature for the beginning of crystallization.

7. Lithium borates showed the greatest tendency toward glass formation. Borates from the oxides MeO increased the tendency toward glass formation with an increase in the atomic weight of Me^{2+} and in the interatomic distances of the corresponding oxides. Thus, while in the $B_2O_3 - Li_2O - BeO$ system, glass formation began only at an oxide ratio $\dfrac{Li_2O}{BeO} > 1$, i.e.,

when lithium borates predominated and the glass mixture crystallized on cooling when beryllium borates predominated, in the $B_2O_3 - Li_2O - CaO$ (BaO, CdO, PbO) systems, the melts crystallized on cooling only for low-alkali glasses ($Li_2O < 2$ mol. %) and glasses containing more than 60 mol. % of B_2O_3. Boric oxide predominated in the crystallization products of these glasses. As a rule, glass formation began as a result of the formation of eutectics and reactions between the solid and liquid phases of different compositions and not just by mechanical solution of the more refractory oxides in a melt of the fusible ones. The early appearance of a liquid phase facilitated borate and glass formation.

8. The general characteristics of the glassy state according to contemporary ideas are also shown by borate glasses. The properties of actual borate glasses, the character of their change at a definite quantitative ratio of the oxides in the glass and also at a change in the ratio $BO_4 : BO_3$ cannot be explained from the point of view of Zachariasen if the structure of the

glass is considered only as an amorphous, continuous, irregular lattice in which the cations of the "modifier" oxides are distributed statistically and the structure of B_2O_3 as a continuous chain of flat $[BO_3]^{3-}$ triangles.

Glass is a cooled melt with a thermodynamically nonequilibrium solid state and a complex composition which are mainly determined by the nature of the oxides and their quantitative content and ratio, as well as by the previous thermochemical treatment of the glass.

9. The results of the series of investigations of the given systems show that binary and more complex borates and other compounds are formed in glasses and there is also a definite amount of unbound B_2O_3, i.e., a definite structure is formed whose foundation was laid during the processes of glass formation. The presence of exothermal effects on the thermograms in the high temperature region, i.e., in the region in which glass formation is completed, confirms the fact that the mass thus undergoes partial ordering of the general structure and equilibration of the state with the evolution of a certain amount of excess free energy.

10. As the investigation shows, the structure formed is deformed to a considerable extent during the completion of glass formation, but is subsequently readily detected during cooling or further thermal and chemical treatment of the glass mixture by the composition of the crystallization products, the results of hydrolytic testing of the glasses, and the character of changes in the physicochemical properties of the glass with a change in temperature, quantitative content, and ratio of the oxides in the glass.

The quantitative condition of the ordered sections and, in connection with this, the character and extent of their effect on the properties of glass require special study both for silicate and other glasses.

11. It was established that a new type of mineral with the composition $3BeO \cdot B_2O_3$ was formed in the $B_2O_3 - Li_2O - BeO$ system and its constants were determined. The optical constants of the other minerals investigated were also determined. In the general case, the composition of the crystallization products of glass (including forced crystallization) showed their direct dependence on the amount and ratio of the oxides $\dfrac{nMeO + mLi_2O}{B_2O_3}$ and also on the

previous thermal treatment of the glass; the crystallization products of the systems investigated were found to contain only those minerals which could be formed at the given ratio of oxides and within the limits of the temperature conditions under which the mixtures were heated. This also confirmed the conclusion presented previously on the structure of borate glasses. The detection of liquation in a series of cases in glasses after their thermal treatment and also the possibility of extracting lithium borates and free B_2O_3 from a series of glasses by hydrolytic treatment emphasizes the complexity of the structure of borate glasses. The results of the investigation may serve as a basis for the preparation of crystallizing glasses with a given structure, i.e., with a high mechanical strength.

12. Analysis of the results of physicochemical investigations of the properties of the glasses shows that the values of these and the character of their changes depend not so much on the quantitative content of the separate oxides as on their quantitative ratio to each other and mainly on the ratio $\dfrac{nMeO + mLi_2O}{B_2O_3}$, i.e., they depend mainly on the structure of the glass,

on the ratio $BO_4 : BO_3$ in the glass, on the degree of binding of B_2O_3 in borates and on the formation of borates in the glass and their nature.

13. Composition-property curves showed more or less sharply expressed extrema at definite ratios $\dfrac{nMeO + mLi_2O}{B_2O_3}$, i.e., in eutectic points, and at the formation of definite borates.

It was established that at first, with an increase in the oxides Li_2O and MeO in the glass up to definite ratio $\dfrac{nMeO + MLi_2O}{B_2O_3}$ there was a definite consolidation and ordering of its structure.

This was shown by the steep rise in the density and refractive-index curves, the decrease in the the thermal expansion of the glass and the sharp fall in its solubility in water. Subsequently, with an increase in the ratio $\dfrac{nMeO + mLi_2O}{B_2O_3}$ there were breaks in the curves, extrema and a corresponding change in the character of the property curves. The disagreement on whether these changes in the properties are due to the conversion of BO_3 into BO_4, i.e., that the extrema occur at a maximum in the conversion of BO_3 to BO_4, or whether these phenomena arise as a result of conversion of BO_4 into BO_3 is not one of principle. Neither side disputes the substance of the changes occurring the glass, which are supported by a great number of experiments. This is the field of B_2O_3 crystallochemistry.

14. The character of the change in the density curves of glasses depends on the composition of the oxides and the ratio of the total oxides, Li_2O and MeO to B_2O_3, while the absolute values of the density which are within the range from 1.9 to 4.5 g/cm^3, depend on the molecular weight of the oxide MeO.

The glass density curves increase sharply until the ratio $\dfrac{nMeO + mLi_2O}{B_2O_3}$ reaches a definite value; then with further increase in this ratio, the density curves undergo a sharp break and become relatively flat.

15. The character of the change in the refractive-index curves of the glass is approximately the same as that of the density curves up to a definite ratio $\dfrac{nMeO + mLi_2O}{B_2O_3}$. When this ratio is exceeded, there are also noticeable breaks in the curves for the increase in N_D, though they are not as sharply expressed and the curves become only somewhat flat. The absolute values of N_D for the glasses investigated lie within the limits of 1.5 to 1.75. The results of investigating N_D of the glasses show that the optical properties of the latter are also "structure-sensitive." The effect of the separate oxides MeO on N_D of the glasses is also proportional to their molecular weights.

A definite relation between the refractive index of the glasses and their density was established.

16. The water resistance of the glasses is directly dependent on the degree of binding of the B_2O_3 in borates, on the ratio $\dfrac{nMeO + mLi_2O}{B_2O_3}$ in the glass and on its density, i.e., mainly on structural factors. The absolute values of the solubility of the glasses in water are from 96 to 0.2%. The highest solubility is shown by glasses in which lithium borates predominate and also high-boron glasses, i.e., glasses in which there is a definite amount of free B_2O_3. The oxides MeO affect the solubility of glass in the following order: the greatest solubility is shown by glasses with the oxides BeO, MgO, and BaO and the least by glasses with the oxides CaO, ZnO, and CdO. The role of the other oxides is intermediate. An increase in the so-called

"oxygen potential" of the glasses, in particular up to the ratio O/B > 1.78 and even up to 1.9 and above, did not produce loosening of the glass structure and an increase in its water solubility, as could be surmised from literature data. The given ratio $\dfrac{nMeO + mLi_2O}{B_2O_3}$ and the density of the glass are the factors which determine its solubility in water. Not only the formation of borates from MeO, but even the formation of lithium borates have an essential effect on the fall in the water solubility of glass. The simultaneous presence of BeO and MgO was found to be very effective in this respect.

17. Analysis of the expansion curves of glasses shows that at definite values of the ratios $\dfrac{Me_2O}{B_2O_3}$ and $\dfrac{mMe_2O \cdot nMeO}{B_2O_3}$ there are sharply expressed extrema, which coincide mainly with those on the density curves of the given glasses. The character of the change in the thermal expansion curves of glasses with a change in composition shows that up to a definite value of the ratios $\dfrac{Me_2O}{B_2O_3}$ and $\dfrac{mMe_2O \cdot nMeO}{B_2O_3}$ compression of the samples occurs, i.e., the heat introduced is consumed mainly in consolidating and ordering the glass structure, but with a further change in the oxide ratio quite rapid expansion occurs.

The thermal expansion of the ternary glass compositions investigated show a slowing down close to their softening point. For glasses with the composition $B_2O_3 - Li_2O - BeO$, the slowing in expansion was found in the temperature range of 320-420°.

The absolute values of $\alpha \cdot 10^{-7}$ of the glasses investigated were within the range from 50 to $110 \cdot 10^{-7}$ and the softening points from 400 to 550°.

18. Depending on the nature of the oxide MeO, the glasses investigated also have the following properties: transparency to x-rays (glasses with BeO and MgO or BeO + MgO), absorption of x-rays (glasses with BaO and PbO), absorption of thermal neutrons (glasses with CdO), etc. These and other properties offer wide prospects for their direct (or with the addition of a series of other oxides, including those of rare earth elements) application in optics, electronics, electrovacuum products, etc.

19. Analysis of the effect of adding from 1 to 70 parts by weight of SiO_2, Al_2O_3, and ZrO_2 to borate glasses also emphasized the complexity of their structure. The addition of up to 50 parts by weight of SiO_2 or Al_2O_3 to borate glasses did not essentially upset the glass melting conditions and produced little change in the behavior of the melts on cooling, but required a certain increase in the melting temperature. Any addition of ZrO_2 led to opaque glass, which is very valuable in the use of the corresponding compositions as enamels, especially for low-melting metals. The addition of SiO_2 and Al_2O_3 to borate glasses did not promote the formation of a stronger structure. The physicochemical properties of the glasses were not essentially improved by the introduction of these additives and sometimes, with the addition of only certain amounts of SiO_2, they were worsened in comparison with the properties of the starting glasses without additives. Together with this, the possibility of introducing up to 50 parts by weight of these additives considerably extends the field of application of these glasses and also of natural borate raw materials without processing and considerably lowers the cost of the glasses.

20. Analysis of the results of determining the properties of glass again emphasizes the error in the practice of developing "universal" coefficients for calculating the properties of

glass, the mechanical application of these obtained for one type of glass to another and also the method of obtaining coefficients for the calculation of the effect of oxides on glass properties just by the simple solution of $\underline{n}$ equations with $\underline{n}$ unknowns. This method "levels" the properties of the oxides and expresses them only as functions of the weight composition, ignoring structural factors, which are the determining ones in this problem. Coefficients expressing the properties of any oxides in the glass do not remain constant. In determining the coefficients for calculating the effect of any oxide on the glass properties, one should start from the values of these for the given oxide in the free state and from the effect of the oxide on the properties of the glass and the changes which the oxide undergoes in the glass. In this respect, the roles of all the oxides are not always equivalent. In borate glasses, the main role as regards the effect on the glass properties and the character of their changes is played by the predominating (and sole acid) oxide, B_2O_3. The roles of the other components are determined largely by the character of their effect on the change in structure of B_2O_3 or the general structure of the glass. For this reason we adopted variable coefficients only for B_2O_3, depending on the system and the ratio $\dfrac{mLi_2O + nMeO}{B_2O_3}$ in the glass.

21. The coefficients for calculating the effect of B_2O_3 on the glass properties in the different glass-forming systems investigated were determined from the following simplified formulas:

$$D = 1.83 + A (100 - B_2O_3) \ g/cm^3;$$
$$V = 38 - A (100 - B_2O_3) \ cm^3/mole;$$
$$N = 1.462 + A (100 - B_2O_3) \text{ for glasses with MeO = BeO, CaO, ZnO, ZrO and}$$
$$\text{BaO;}$$
$$N = 1.462 + A \cdot MeO \text{ for glasses with MeO = MgO, CdO and PbO.}$$

For calculating $\alpha \cdot 10^{-7}{}_{B_2O_3}$ it was necessary to use complex formulas of the type $Y + C = A(X - B)^3$ (an equation of the cubic parabolic type) or $Y = Ae^{-BX} - 152$ (a complex power function). Satisfactory results were also given by the following simplified formula: $Y = 1.52 \cdot B_2O_3 - C$.

22. It can be assumed that under certain conditions the cations of the oxides MeO, especially Be^{2+}, Cd^{2+}, and Pb^{2+}, sometimes participate in the organization of the glass structure together with boron. However, it is not worthwhile having differentiated coefficients for these oxides for calculating their effect on the glass properties, as in the over-all result the roles of all the oxides MeO and the degrees of their effect on the structure of these borate glasses are not very different. The oxide which has a determining effect on all the glass properties is B_2O_3 alone.

23. Composition — property diagrams were constructed characterizing the effect of the separate oxides on the properties of the glass and also diagrams of glass formation, density, refractive index, water solubility, and thermal expansion for all the systems investigated; corresponding diagrams were also given for glasses with the composition $B_2O_3 - Li_2O - BeO + MgO$ and also photoreproductions of microsections of forcibly crystallized glasses, thermograms, and weight-loss curves.

CONCLUSION

The results of the series of investigations and the scientific generalization of problems in the thermochemistry, structure, and physical and physicochemical properties of borate glasses, the rules of their changes and calculation methods, the thermotechnics and technology of melting, working, and annealing these glasses, chemical methods for analyzing their composition, and also the mineralogy and crystallooptics of the crystallization products serve as a basis:

a) for a scientific solution of problems in the thermochemistry and reaction kinetics of borate formation, the rules of changes in the properties of borate glasses and also methods of calculating them;

b) for a scientific synthesis of new compositions for glasses and enamels for different purposes and with definite, given properties: clear and opaque; crystallizing with micro- and macrocrystallites of definite composition, dimensions and mutual disposition in the glass mass; with selective transparency to or adsorption of long- or short-wave rays and thermal neutrons and also shielding against various ultrashort-wave rays; with high water resistance, not less than that of optimal compositions of silicate and borosilicate glasses, or, on the other hand, with almost complete decomposition and solution when treated for a definite time with moisture and various acid reagents; with a density in the range from 1.9 to 4.5 g/cm^3; with a refractive index within the range from 1.5 to 1.75 and above; with a coefficient of linear thermal expansion in the range from 40 to $110 \cdot 10^{-7}$ and above, etc.

All this considerably broadens the region of possible use of borate glasses as electrovacuum, electrotechnical and optical glasses, glasses with a high microhardness, glasses for the manufacture of different glass fibers, enamels for various metals, etc.

c) for a fundamental improvement in the compositions, technology and properties of workable borate glasses, as has already been proved under actual factory conditions;

d) for solving the problems of using various natural borate, lithium and other raw materials in the production of glasses and enamels without processing, considering the silicates, aluminosilicates, etc., present in them;

e) for further developments in the scientific investigations of other polycomponent borate systems and in the scientific and technological problems mentioned above.

TABLES OF THE COMPOSITIONS OF GLASSES AND THEIR PHYSICOCHEMICAL PROPERTIES

TABLE 1

B_2O_3 — Monocomponent System, According to Karkhanavala [166]

Sample No.	B_2O_3, %	$\alpha \cdot 10^{-7}$ experimental	Temperature range, °C	Investigators
1	100	151,0	—	Cousen and Turner (1938)
2	"	152,7	0—100	Gooding and Turner (1934)
3	"	141,4	0—100	Grenet (1896)
4	"	144,6	0—100	Grenet (1897)
5	"	143,0	about 0	Samsoen (1925)
6	"	151,8	about 100	Samsoen (1926)
7	"	143,0	about 0	Samsoen (1929)
8	"	145,0	about 100	Samsoen (1929)
9	"	154,1	0—100	Seddon and Turner (1933)

Unannealed glass

1	100	133,1	about 122,1	
2	"	63,3	" 152,3	
3	"	— 63,3	" 180,4	
4	"	— 356,3	" 204,2	Spag and Park (1954)
5	"	183,0	" 229,4	
6	"	1366,0	" 250,2	
7	"	2073,3	" 270,0	

Well annealed glass

1	100	160,0	about 114,2	
2	"	122,5	" 143,4	
3	"	147,5	" 171,8	
4	"	393,3	" 199,6	Spag and Park (1954)
5	"	1350,0	" 226,9	
6	"	2043,3	" 250,0	
7	"	1973,0	" 270,0	
1	100	152,0	0—100	Weinig and Zschimmer (1929)

TABLE 2

Glasses with the Composition $Li_2O-B_2O_3$

a) according to Bresker and Evstrop'ev [59]

Sample No.	Composition, mol.% B$_2$O$_3$	Li$_2$O	$\frac{Li_2O}{B_2O_3}$	$\frac{O}{B}$	D of glass in g/cm³	V$_{B_2O_3}$ in cm³/mole	N$_D$ of glass	N$_D$ B$_2$O$_3$
1	100,0	—	—	1,5	1,8343	38,0	1,4625	1,4625
2	97,87	2,13	0,021	1,515	1,8621	37,6	1,4687	1,4650
3	95,86	4,14	0,043	1,525	1,8827	37,2	1,4723	1,4720
4	93,82	6,18	0,066	1,535	1,9145	36,6	1,4810	1,472
5	91,84	8,16	0,089	1,545	1,9457	36,0	1,4867	1,475
6	89,77	10,23	0,114	1,56	1,9772	35,8	1,4937	1,480
7	87,76	12,24	0,14	1,57	2,0094	35,4	1,5009	1,487
8	85,79	14,21	0,167	1,58	2,0381	34,7	1,5067	1,495
9	83,65	16,35	0,196	1,595	2,0682	34,5	1,5136	1,499
10	81,84	18,16	0,22	1,605	2,0941	34,1	1,5188	1,503
11	79,73	20,27	0,25	1,63	2,1228	33,7	1,5221	1,506
12	77,74	22,26	0,29	1,64	2,1506	33,3	1,5307	1,508
13	75,91	24,09	0,32	1,66	2,1768	33,0	1,5364	1,520
14	74,02	25,98	0,35	1,67	2,1999	32,5	1,5441	1,596

b) according to Karkhanavala [166]

Sample No.	Composition, mol.% B$_2$O$_3$	Li$_2$O	$\frac{Li_2O}{B_2O_3}$	$\frac{O}{B}$	$\alpha \cdot 10^{-7}$ of glass	$\alpha \cdot 10^{-7}$ temperature range °C	$\alpha \cdot 10^{-7}$ B$_2$O$_3$	Investigators
1	90,3	9,7	0,11	1,55	90,0	18—100	70,6	Grenet(1896)
2	83,3	16,7	0,2	1,59	64,6	"	23,5	
3	80,0	20,0	0,25	1,62	59,0	"	6,25	
4	75,0	25,0	0,33	1,67	67,8	"	0,4	
1	90,3	9,7	0,11	1,55	93,4	"	74,0	Grenet (1897)
2	83,3	16,7	0,2	1,59	67,8	"	27,5	
3	80,0	20,0	0,25	1,62	62,2	"	10,2	
4	75,0	25,0	0,33	1,67	71,0	"	4,7	

Note: In the present and subsequent tables in which the data from investigations of other authors are presented, we calculated the values of $\frac{MeO}{B_2O_3}$, $\frac{Me_2O}{B_2O_3}$, $\frac{MeO + Me_2O}{B_2O_3}$, $\frac{O}{B}$, $V_{B_2O_3}$, $N_{B_2O_3}$ and $\alpha_{B_2O_3}$. Where the composition of the glass was given in wt.%, we present it in mol.%.

T A B L E 3

Glasses with the Composition Na_2O – B_2O_3, According to Abe [92]

Sample No.	Composition, mol. %		$\dfrac{Na_2O}{B_2O_3}$	$\dfrac{O}{B}$	D of glass in g/cm³	$V_{B_2O_3}$ in cm³/mole	N_D		Investigators
	B_2O_3	Na_2O					of glass	B_2O_3	
1	100,0	—	—	1,5	1,812	38,4	1,4582	1,4582	Wulff and Majumdar
2	98,0	2,0	0,021	1,51	1,8662	37,9	—	—	Gooding and Turner
3	96,5	3,5	0,036	1,52	1,8632	38,4	—	—	
4	92,7	7,3	0,079	1,54	1,9876	35,9	1,4841	1,475	Wulff and Majumdar
5	91,3	8,7	0,095	1,55	1,9637	37,1	—	—	Gooding and Turner
6	87,5	12,5	0,143	1,57	2,0631	35,2	—	—	
7	85,0	15,0	0,177	1,59	2,1295	34,3	—	—	
8	82,5	17,5	0,212	1,61	2,1369	34,4	—	—	
9	80,5	19,5	0,242	1,63	—	—	1,501	1,476	Morey and Merwin
10	79,8	20,2	0,254	1,63	2,2309	33,3	—	—	Gooding and Turner
11	76,5	23,5	0,307	1,66	2,2436	33,3	—	—	
12	68,0	32,0	0,47	1,73	2,3703	31,1	—	—	
13	66,5	33,5	0,505	1,75	2,370	32,3	—	—	Tillotson
14	66,5	33,5	0,505	1,75	—	—	1,515	1,478	Bedson

TABLE 3 (continued)

Sample No.	Composition, mol.% B₂O₃	Composition, mol.% Na₂O	$\frac{Na_2O}{B_2O_3}$	$\frac{O}{B}$	D of glass in g/cm³	$V_{B_2O_3}$ in cm³/mole	N_D of glass	N_D B₂O₃	Investigators
15	93,0	7,0	0,075	1,53	1,92	37,1	—	—	Stevels
16	89,0	11,0	0,124	1,50	2,02	35,7	—	—	
17	82,0	18,0	0,22	1,61	2,16	33,9	—	—	
18	78,0	22,0	0,28	1,64	2,23	33,3	—	—	
19	68,5	31,5	0,46	1,73	2,36	31,0	—	—	
				Annealed glasses [92]					
1	100	—	—	1,50	1,812	38,4	1,458	1,458	
2	94,39	5,61	0,066	1,53	1,912	37,0	1,479	1,480	
3	88,92	11,08	0,139	1,57	2,024	35,5	1,489	1,478	
4	83,46	16,54	0,222	1,61	2,140	33,8	1,4965	1,480	
5	78,06	21,94	0,313	1,65	2,237	32,6	1,5035	1,480	
6	72,75	27,23	0,42	1,71	2,318	31,8	1,501	1,465	
7	67,49	32,51	0,54	1,77	2,369	31,4	1,515	1,490	
				Quenched glasses [92]					
1	100	—	—	1,5	1,778	39,1	1,4502	1,4502	
2	92,38	7,62	0,091	1,54	1,969	36,1	1,4797	1,470	
3	89,84	10,16	0,125	1,57	2,009	35,9	1,4839	1,472	
4a	81,82	18,18	0,246	1,62	2,1226	34,3	1,4953	1,474	
4b	81,82	18,18	0,246	1,62	2,1235	34,3	1,4958	1,475	
5	74,67	25,33	0,38	1,69	2,2305	32,1	1,5051	1,475	
6	69,3	30,70	0,49	1,74	2,3285	31,7	1,5136	1,475	
7a	69,17	30,83	0,49	1,75	2,3322	31,7	1,5144	1,480	
7b	69,17	30,83	0,49	1,75	2,3324	31,7	1,5147	1,480	

TABLE 3 (continued)

Glasses with the composition Na$_2$O – B$_2$O$_3$, according to Karkhanavala [166]

Sample No.	Composition, mol.%		Na$_2$O / B$_2$O$_3$	O / B	α · 10^{-7}			Investigators
	B$_2$O$_3$	Na$_2$O			of glass	temperature range, °C	B$_2$O$_3$	
1	97,5	2,5	0,039	1,51	141,2	0—100	134,0	
2	96,0	4,0	0,042	1,52	138,6	"	128,0	
3	94,4	5,6	0,06	1,53	127,0	"	111,0	
4	91,3	8,7	0,095	1,54	110,2	"	83,0	
5	88,9	11,1	0,125	1,56	100,0	"	63,0	Gooding and Turner (1934)
6	87,3	12,7	0,145	1,575	93,2	"	49,5	
7	82,6	17,4	0,21	1,61	84,9	"	20,0	
8	79,6	20,4	0,26	1,63	89,5	"	11,3	
9	78,0	22,0	0,28	1,64	90,0	"	3,9	
10	76,0	24,0	0,32	1,66	91,3	"	—4,2	
11	73,0	27,0	0,37	1,685	96,1	"	—13,7	
12	67,4	32,6	0,49	1,74	102,5	"	—40,0	
1	94,4	5,6	0,06	1,53	101,9	18—100	85,0	
2	91,0	9,0	0,099	1,55	93,3	"	64,0	
3	83,6	16,4	0,2	1,6	81,7	"	20,2	Grenet (1896)
4	71,5	28,5	0,4	1,7	70,4	"	—59,0	
5	66,6	33,4	0,5	1,75	95,3	"	—55,0	

TABLE 3 (continued).

Sample No.	Composition, mol.%		Na₂O/B₂O₃	O/B	α·10⁻⁷			Investigators
	B₂O₃	Na₂O	$\frac{Na_2O}{B_2O_3}$	$\frac{O}{B}$	of glass	tempera-ture range °C	B₂O₃	
1	94,4	5,6	0,06	1,53	105,0	18—100	88,0	
2	91,0	9,0	0,099	1,55	96,7	"	67,0	
3	83,8	16,2	0,2	1,6	87,6	"	28,0	Grenet (1935)
4	69,0	31,0	0,45	1,72	94,2	"	—41,0	
5	66,8	33,2	0,5	1,75	111,9	"	—28,5	
1	99,1	0,9	0,009	1,51	132,0	0— 100	129,0	
2	98,4	1,6	0,016	1,51	129,0	"	124,0	
3	94,4	5,6	0,06	1,53	103,0	"	87,0	
4	92,5	7,5	0,08	1,54	98,5	"	74,5	
5	91,1	8,9	0,098	1,55	95,0	"	66,0	Samsoen (1928)
6	83,5	16,5	0,2	1,6	86,4	"	26,0	
7	75,8	24,2	0,82	1,65	93,0	"	—3,4	
8	66,7	33,3	0,5	1,75	111,0	"	—31,0	
9	64,0	36,0	0,56	1,78	106,0	"	—56,0	
10	55,8	44,2	0,79	1,92	80,5	"	—167,0	

TABLE 4

Glasses with the Composition $K_2O - B_2O_3$ [166]

Sample No.	Composition, mol. %		$\dfrac{K_2O}{B_2O_3}$	$\dfrac{O}{B}$	$\alpha \cdot 10^{-7}$			Investigators
	B_2O_3	K_2O			of glass	temperature range, °C	B_2O_3	
1	100	—	—	1,5	159,3	0—100°	159,3	Grenet (1942)
2	91,8	8,2	0,085	1,54	135,0	.	105,0	
3	88,5	11,5	0,13	1,57	121,4	.	77,0	
4	83,4	16,6	0,20	1,59	117,7	.	49,0	
5	80,0	20,0	0,25	1,62	117,5	.	30,0	
6	75,5	24,5	0,33	1,66	124,9	.	14,5	
7	66,5	33,5	0,50	1,75	160,3	.	6,5	

TABLE 5

Glasses with the Composition $PbO - B_2O_3$ [166]

Sample No.	Composition, mol. %		$\dfrac{PbO}{B_2O_3}$	$\dfrac{O}{B}$	$\alpha \cdot 10^{-7}$			Investigators
	B_2O_3	PbO			of glass	temperature range, °C	B_2O_3	
1	75,2	24,8	0,33	1,66	69,5	0—100°	40,0	Samsoen (1938)
2	72,0	28,0	0,39	1,69	63,0	.	25,0	
3	67,0	33,0	0,49	1,75	69,5	.	25,0	
4	50,0	50,0	1,00	2,00	99,0	.	38,0	
5	40,0	60,0	1,50	2,24	11,0	.	37,5	
6	33,5	66,5	2,00	2,48	24,0	.	52,0	
7	24,5	75,5	3,10	3,00	29,5	.	35,0	

TABLE 6

Glasses with the Composition $B_2O_3 - Na_2O - BeO$ [70]

Sample No.	Composition, mol. %			$\dfrac{Na_2O + BeO}{B_2O_3}$	$\dfrac{O}{B}$	D of glass, g/cm³	$V_{B_2O_3}'$, cm³/mole	N_D	
	B_2O_3	Na_2O	BeO					of glass	B_2O_3
1	50,49	32,29	17,21	0,80	1,98	2,3502	34,5	1,5108	1,4300
2	54,29	30,40	15,20	0,84	1,92	2,3477	34,0	1,5122	1,4435
3	59,41	30,53	10,06	0,68	1,83	2,3494	33,2	1,5118	1,4575
4	64,05	30,41	5,54	0,57	1,78	2,3463	32,7	1,5121	1,4680
5	44,18	39,46	26,37	1,26	2,13	2,3408	35,3	1,5135	1,4115
6	64,19	20,51	15,30	0,56	1,77	2,2558	34,0	1,5080	1,4611
7	59,05	25,87	15,07	0,70	1,85	2,3154	33,5	1,5125	1,4571
8	59,13	20,61	20,76	0,70	1,84	2,2734	34,3	1,5134	1,4587
9	49,66	25,03	25,31	1,01	2,00	2,3228	34,6	1,5163	1,4391
10	45,78	24,96	29,75	1,16	2,11	2,3285	35,2	1,5185	1,4233
11	54,17	25,64	20,19	0,85	1,93	2,3149	34,3	1,5121	1,4443
12	64,81	24,82	10,37	0,54	1,77	2,2960	33,5	1,5124	1,4695
13	69,01	20,85	10,13	0,45	1,73	2,2339	35,0	1,5052	1,4655
14	46,76	30,27	19,98	1,14	2,02	2,3383	35,0	1,5130	1,4290
15	46,76	29,51	24,22	1,14	2,07	2,3392	35,4	1,5147	1,4250
16	51,46	24,43	24,14	0,94	1,97	2,3183	34,3	1,5156	1,4431
17	40,67	29,60	29,73	1,45	2,24	2,3444	36,2	1,5140	1,4000
18	72,03	22,73	5,24	0,39	1,69	2,2518	33,3	1,5062	1,4733
19	84,10	14,00	1,90	0,19	1,59	2,1449	34,0	1,4973	1,4792
20	81,52	15,72	2,76	0,23	1,61	2,1410	34,4	1,4955	1,4739
21	87,62	9,87	2,51	0,14	1,57	2,0547	35,1	1,4911	1,4770
22	67,87	32,12	1,63	0,47	1,75	2,3627	32,7	1,5145	1,4700

TABLE 7

Glasses with the Composition $B_2O_3 - Na_2O - BeO$ [155]

Glass No.	Composition, mol. %			$\frac{Na_2O+BeO}{B_2O_3}$	$\frac{O}{B}$	D of glass, g/cm³	$V_{B_2O_3}$, cm³/mole	$\alpha \cdot 10^{-7}$ of glass	temperature range, °C	$\alpha \cdot 10^{-7}$ B_2O_3
	B_2O_3	Na_2O	BeO							
70	75	15	10	0,33	1,68	2,178	34,4	98,0	20—400	46,0
71	70	20	10	0,43	1,71	2,194	34,6	95,1	.	16,5
72	65	25	10	0,54	1,77	2,231	34,4	98,9	.	—7,1
73	60	30	10	0,66	1,83	2,298	34,4	117,9	.	—7,7
74	55	35	10	0,82	1,91	2,344	34,1	113,7	.	—52,5
75	85	10	5	0,18	1,59	—	—	95,6	.	—63,5

TABLE 8

Glasses with the Composition $B_2O_3 - K_2O - BeO$ [77]

Glass No.	Composition, mol. %			$\frac{K_2O+BeO}{B_2O_3}$	$\frac{O}{B}$	D of glass, g/cm³	$V_{B_2O_3}$, cm³/mole	N_D of glass	N_D B_2O_3
	B_2O_3	K_2O	BeO						
156	43,02	29,26	27,71	1,32	2,16	2,2673	37,8	—	—
158	66,43	28,20	5,45	0,50	1,75	2,2554	34,1	1,4919	1,448
165	63,93	31,58	4,49	0,56	1,785	2,2919	33,8	1,4975	1,450
166	55,69	30,17	14,14	0,80	1,895	2,2717	35,4	—	—
169	71,77	18,04	10,20	0,40	1,695	2,1610	35,2	1,4866	1,450
173	53,48	24,57	21,96	0,85	1,935	2,2392	36,4	1,4972	1,425
174	57,62	24,29	18,08	0,72	1,865	2,2337	35,8	1,4938	1,430
175	52,69	28,97	18,34	0,90	1,950	2,2672	35,9	1,4998	1,430
176	67,82	18,09	14,09	0,47	1,740	2,1797	33,2	1,4914	1,450
177	61,90	23,99	14,40	0,68	1,810	2,2279	25,0	1,4939	1,440
178	65,75	24,06	10,16	0,52	1,760	2,2234	34,6	1,4926	1,445
179	59,19	30,28	10,53	0,67	1,810	2,2761	35,0	1,5120	1,450
180	77,35	17,95	4,69	0,28	1,650	2,1406	35,2	1,4838	1,460
181	47,92	25,35	26,73	1,10	2,040	2,2547	36,6	1,4994	1,410
182	48,75	29,02	22,23	1,05	2,020	2,2682	36,4	—	—
184	71,75	23,50	4,69	0,33	1,690	2,2090	34,6	1,4886	1,455

TABLE 9

Glasses with the Composition $B_2O_3 - Li_2O - BeO$ [61]

| Glass No. | Composition, mol. % | | | $\dfrac{Li_2O+BeO}{B_2O_3}$ | $\dfrac{O}{B}$ | D of glass g/cm³ |
	B_2O_3	Li_2O	BeO			
13	47,0	26,0	27,0	1,14	1,98	2,2754
14	47,5	24,5	28,0	1,10	2,00	2,2719
15	46,0	27,5	26,5	1,17	2,00	2,2793
20	54,0	23,5	22,5	0,85	1,87	2,1892
21	51,0	27,0	22,0	0,96	1,97	2,2429
22	49,0	29,0	22,0	1,04	1,93	2,2599
26	62,0	19,0	19,0	0,61	1,78	2,0998
27	59,0	23,5	17,5	0,70	1,81	2,1596
28	56,5	26,0	17,5	0,77	1,86	2,1973
29	52,5	30,0	17,5	0,90	1,91	2,2564
30	52,5	30,5	17,5	0,90	2,00	2,2546
33	71,0	15,5	13,5	0,41	1,68	2,0677
34	68,5	20,0	11,5	0,46	1,73	2,0870
35	63,5	23,5	13,0	0,58	1,76	2,1159
36	60,0	27,4	12,6	0,67	1,77	2,1077
37	57,0	31,0	12,0	0,76	1,80	2,1478
38	48,5	40,0	11,5	1,06	1,85	2,2092
39	53,5	34,0	12,5	0,87	1,90	2,2122
40	51,0	37,6	11,4	0,96	1,95	2,2743
41	63,5	25,0	11,5	0,58	2,00	2,2850
42	78,0	12,6	9,4	0,28	1,63	2,0643
43	72,5	20,2	7,3	0,38	1,67	2,0766
44	66,2	27,0	6,8	0,52	1,71	2,0943
45	65,5	27,5	7,0	0,53	1,74	2,1073
46	65,0	28,3	6,7	0,54	1,75	2,1132
G-1	50,5	24,3	25,2	0,98	2,00	2,2200
G-2	50,2	33,1	16,7	0,99	2,00	2,2600
Schleede and Wellmann	65,2	25,1	9,7	0,53	1,77	2,0900
Lindemann	63,2	30,6	6,2	0,58	1,75	2,0900

Glasses with the Composition $B_2O_3 - Li_2O - BeO$ [155]

Glass No.	B_2O_3	Li_2O	BeO	$\dfrac{Li_2O+BeO}{B_2O_3}$	$\dfrac{O}{B}$	D of glass g/cm³
65	75,0	15,0	10,0	0,33	1,67	—
66	70,0	20,0	10,0	0,43	1,71	2,161
67	65,0	25,0	10,0	0,54	1,76	2,212
68	60,0	30,0	10,0	0,66	1,83	2,244
69	57,5	32,5	10,0	0,73	1,86	—

TABLE 9 (continued)

$V_{B_2O_3}$, cm³/mole	N_D		$\alpha \cdot 10^{-7}$			Solubility in water, %
	of glass, exper.	of B_2O_3, calc.	of glass, exper.	temperature range, °C	of B_2O_3, calc.	
33,3	1,562	1,470	89,1	20—420	21,3	40,7
33,4	1,558	1,465	—	"	—	36,3
34,4	1,564	1,490	93,4	"	22,3	43,0
34,2	1,552	1,470	74,2	"	6,53	60,9
33,3	1,558	1,470	81,2	"	—1,4	53,8
33,4	1,564	1,470	89,6	"	8,0	48,5
35,0	1,542	1,480	71,8	"	25,0	75,1
34,2	1,550	1,480	72,0	"	5,5	76,3
34,0	1,550	1,470	75,8	"	3,8	77,4
33,0	1,562	1,475	80,0	"	—11,2	80,7
33,5	1,564	1,495	83,3	"	— 1,2	80,1
35,0	1,530	1,480	62,6	"	23,4	80,3
35,0	1,538	1,480	67,6	"	15,5	86,8
34,8	1,548	1,485	71,3	"	9,9	88,9
34,0	1,549	1,470	71,5	"	— 7,0	91,3
34,3	1,552	1,465	73,5	"	—22,0	88,6
33,8	1,560	1,440	76,9	"	—65,3	88,6
33,4	1,566	1,475	88,6	"	—12,1	89,8
33,2	1,570	1,475	94,0	"	—16,4	89,9
33,2	1,574	1,520	101,7	"	53,0	90,3
36,0	1,520	1,480	—	"	—	88,6
35,0	1,536	1,485	63,1	"	9,7	91,8
34,5	1,547	1,480	69,0	"	6,3	92,4
34,7	1,549	1,495	66,7	"	— 6,3	92,6
34,8	1,550	1,485	69,5	"	— 6,0	93,6
34,5	1,556	1,470	—	—	—	48,0
33,5	1,566	1,480	—	—	—	72,0
35,0	1,545	1,470	—	—	—	90,0
34,8	1,546	1,477	—	—	—	92,0
—	—	—	65,6	20—400	27,5	—
33,6	—	—	68,1	"	13,7	—
33,0	—	—	75,7	"	5,4	—
33,0	—	—	86,1	"	1,0	—
—	—	—	89,7	"	— 4,9	—

TABLE 10

Glasses with the Composition $B_2O_3 - Li_2O - MgO$ [77]

Glass No.	Composition, mol. %			$\dfrac{Me_2O + MgO}{B_2O_3}$	$\dfrac{O}{B}$	D of glass, g/cm^3	$V_{B_2O_3}$, cm^3/mole	N_D	
	B_2O_3	Me_2O	MgO					of glass	B_2O_3
263	79,65	12,34	8,01	0,25	1,63	2,1567	32,2	1,528	1,49 5
265	76,23	12,10	11,67	0,32	1,63	2,2119	32,9	1,533	1,495
268	72,85	12,23	14,92	0,37	1,69	2,2619	32,3	1,543	1,505
269	68,50	12,44	19,06	0,46	1,73	2,3196	31,6	1,5505	1,509
286	83,71	11,84	4,45	0,19	1,60	2,0871	34,1	1,5142	1,470
289	78,47	17,00	4,53	0,275	1,63	2,1607	34,1	1,5274	1,470
291	75,91	15,98	8,11	0,315	1,66	2,2040	34,0	1,5348	1,495
292	72,09	15,97	11,94	0,39	1,69	2,2619	32,0	1,5441	1,455
293	68,80	15,87	15,33	0,45	1,72	2,3103	32,5	1,5516	1,435
296	76,00	19,95	4,05	0,32	1,66	2,1907	33,3	1,5348	1,475

Glasses with the composition $B_2O_3 - Na_2O - MgO$ [77]

Glass No.	B_2O_3	Me_2O	MgO	$\dfrac{Me_2O+MgO}{B_2O_3}$	$\dfrac{O}{B}$	D of glass, g/cm^3	$V_{B_2O_3}$	of glass	B_2O_3
300	79,20	9,59	11,21	0,26	1,63	2,2440	32,8	1,5143	1,495
305	82,46	7,86	9,68	0,21	1,60	2,1272	34,2	1,5042	1,490
307	72,27	16,64	11,09	0,30	1,69	2,3041	34,8	1,5197	1,490
310	73,70	20,68	5,62	0,36	1,68	2,2866	32,6	1,5134	1,485
313	88,27	5,82	5,91	0,145	1,57	2,0671	34,8	1,4952	1,482
314	83,42	10,97	5,61	0,20	1,60	2,1429	34,0	1,5020	1,490
315	78,19	16,12	5,69	0,27	1,63	2,2321	34,6	1,5034	1,48

Glasses with the composition $B_2O_3 - K_2O - MgO$ [77]

Glass No.	B_2O_3	Me_2O	MgO	$\dfrac{Me_2O+MgO}{B_2O_3}$	$\dfrac{O}{B}$	D of glass, g/cm^3	$V_{B_2O_3}$	of glass	B_2O_3
335	55,51	26,57	17,92	0,80	1,90	2,2391	36,9	1,4914	1,415
341	64,36	27,30	8,34	0,56	1,78	2,2608	31,2	1,4969	1,455
347	69,62	17,72	12,66	0,44	1,72	2,2566	33,1	1,5058	1,470
348	59,64	27,71	12,65	0,68	1,78	2,2490	35,6	1,4991	1,445
349	74,45	17,46	8,09	0,345	1,68	2,2147	34,0	1,4969	1,470
350	83,58	8,02	8,40	0,20	1,60	2,1091	34,9	1,4961	1,480
351	72,80	22,66	4,54	0,375	1,69	2,2289	34,6	1,4987	1,476

TABLE 11

Glasses with the Composition $B_2O_3 - BeO - MgO$ [155]

Glass No.	Composition, mol. %			$\dfrac{BeO+MgO}{B_2O_3}$	$\dfrac{O}{B}$	D of glass, g/cm^3	$V_{B_2O_3}$, cm^3/mole	$\alpha \cdot 10^{-7}$	
	B_2O_3	BeO	MgO					of glass, temperature range, °C	B_2O_3
84	45,0	20,0	35,0	1,22	2,12	2,405	33,8	52,6	20 — 400	50,0
85	40,0	25,0	35,0	1,50	2,25	—	—	55,7	"	59,0
86	45,0	15,0	40,0	1,22	2,11	2,441	33,3	45,6	"	40,0
87	40,0	20,0	40,0	1,50	2,25	2,452	33,6	55,0	"	55,0

TABLE 12

Glasses with the Composition $B_2O_3 - Li_2O - MgO$ [155]

Glass No.	Composition, mol.%			$\dfrac{Me_2O+MgO}{B_2O_3}$	$\dfrac{O}{B}$	D of glass, g/cm³	$V_{B_2O_3}$, cm³/mole	$\alpha \cdot 10^{-7}$		
	B_2O_3	Me_2O	MgO					of glass	temperature range, °C	B_2O_3
33	80,0	10,0	10,0	0,25	1,62	2,117	34,0	63,5	20—400	38,0
34	75,0	15,0	10,0	0,33	1,67	2,195	33,3	67,8	"	28,5
35	70,0	20,0	10,0	0,43	1,71	2,278	32,1	67,8	"	11,0
36	67,5	22,5	10,0	0,48	1,74	—	—	70,5	"	5,2
37	90,0	10,0	—	0,11	1,56	—	—	93,2	"	73,5
38	85,0	10,0	5,0	0,18	1,59	2,068	34,6	72,8	"	50,5
39	82,5	10,0	7,5	0,21	1,61	—	—	66,0	"	42,5
40	75,0	10,0	15,0	0,33	1,67	2,271	34,7	63,0	"	43,0
41	70,0	10,0	20,0	0,43	1,71	—	—	61,2	"	36,3
42	80,0	20,0	—	0,25	1,63	—	—	73,1	"	24,0
43	75,0	20,0	5,0	0,33	1,66	—	—	68,0	"	14,7
44	65,0	20,0	15,0	0,54	1,77	—	—	70,0	"	9,3
45	60,0	20,0	20,0	0,67	1,83	2,378	31,5	70,8	"	8,0
46	55,0	20,0	25,0	0,82	1,92	—	—	73,1	"	7,5

TABLE 12 (continued)

Glass No.	Composition, mol. %			$\dfrac{Me_2O+MgO}{B_2O_3}$	$\dfrac{O}{B}$	D of glass, g/cm^3	$V_{B_2O_3}$, cm^3/mole	$\alpha \cdot 10^{-7}$		
	B_2O_3	Me_2O	MgO					of glass	temperature range, °C	B_2O_3
47	85,0	5,0	10,0	0,18	1,61	2,089	34,5	74,1	20—400	65,0
48	80,0	10,0	10,0	0,25	1,63	2,152	34,5	75,5	•	40,0
49	75,0	15,0	10,0	0,33	1,67	2,235	34,2	80,7	•	19,5
50	70,0	20,0	10,0	0,43	1,71	2,293	33,0	88,0	•	4,3
51	65,0	25,0	10,0	0,54	1,76	—	—	104,0	•	— 0,5
52	60,0	30,0	10,0	0,67	1,83	2,396	32,6	120,5	•	— 3,3
53	90,0	10,0	—	0,11	1,55	—	—	103,1	•	62,0
54	85,0	10,0	5,0	0,18	1,61	2,084	35,0	91,0	•	53,0
55	82,5	10,0	7,5	0,21	1,60	—	—	79,5	•	45,0
56	75,0	10,0	15,0	0,33	1,67	2,213	33,6	74,5	•	40,0
57	70,0	10,0	20,0	0,43	1,71	2,281	33,2	73,7	•	30,0
58	65,0	10,0	25,0	0,54	1,77	2,336	32,8	73,9	•	25,0
59	75,0	25,0	—	0,33	1,67	—	—	100,8	•	2,8
60	75,0	20,0	5,0	0,33	1,67	—	—	92,5	•	11,0
61	75,0	18,0	7,0	0,33	1,67	—	—	86,8	•	13,5
62	65,0	20,0	15,0	0,54	1,76	—	—	91,7	•	4,0
63	65,0	22,5	12,5	0,54	1,77	—	—	97,5	•	1,5
64	50,0	30,0	20,0	1,00	1,99	—	—	111,0	•	—18,0

TABLE 13

Glasses with the Composition $B_2O_3 - Li_2O - MgO$ [62]

Glass No.	Composition, mol.%			$\dfrac{Li_2O+MgO}{B_2O_3}$	$\dfrac{O}{B}$	D of glass, g/cm³	$V_{B_2O_3}$', cm³/mole	N_D		Solubility in water, %	X-ray transparency, % (t = 1,0 cm)
	B_2O_3	Li_2O	MgO					of glass	B_2O_3		
9	80,90	5,16	13,94	0,24	1,62	2,1807	33,1	1,519	1,495	21,9	73,8
10	79,70	5,00	15,30	0,255	1,62	2,1911	33,2	1,522	1,495	14,4	73,7
13	85,55	9,69	4,76	0,22	1,58	2,0503	34,7	—	—	96,4	75,3
14	83,22	10,51	6,27	0,20	1,595	2,1324	33,5	1,519	1,490	96,4	74,3
15	82,94	10,17	6,89	0,21	1,60	2,1560	33,3	1,520	1,490	71,7	74,2
16	80,72	10,29	8,99	0,24	1,62	2,1600	33,4	1,526	1,495	32,9	74,1
17	79,01	10,30	10,60	0,27	1,63	2,1853	32,2	1,528	1,500	19,5	73,8
18	77,40	10,22	12,38	0,29	1,64	2,2070	32,9	1,531	1,495	16,3	73,6
19	76,05	10,02	13,93	0,32	1,65	2,2306	32,1	1,532	1,495	15,4	73,3
20	75,10	10,15	14,75	0,335	1,67	2,2436	32,3	1,538	1,500	9,2	73,2
21	82,82	15,80	1,58	0,21	1,60	2,0660	33,2	1,522	1,490	95,5	75,2
22	81,43	15,65	2,92	0,23	1,61	2,1334	33,7	1,522	1,525	85,7	74,5
23	80,20	15,41	4,39	0,25	1,62	2,1612	33,2	1,525	1,560	55,8	74,2
24	78,85	15,23	5,92	0,265	1,635	2,1680	33,0	1,530	1,490	42,0	74,1
25	77,56	14,56	7,88	0,29	1,645	2,1823	33,0	1,530	1,500	26,0	73,9
26	76,41	14,57	9,02	0,31	1,65	2,2010	32,7	1,537	1,500	18,2	73,7
27	74,80	14,78	10,42	0,34	1,67	2,2355	32,7	1,540	1,495	21,1	73,3
28	73,25	14,93	11,82	0,365	1,685	2,2482	32,2	1,545	1,530	18,8	73,1
29	72,69	14,65	12,66	0,375	1,69	2,3534	31,0	1,510	1,495	15,9	72,1
30	70,95	14,65	14,40	0,41	1,70	2,3654	31,0	1,546	1,500	14,5	72,0
31	78,42	20,00	1,58	0,276	1,64	2,1420	33,2	1,532	1,490	79,2	74,9
32	76,95	19,95	3,10	0,30	1,65	2,1714	33,5	1,537	1,495	53,7	74,1
33	76,07	19,65	4,28	0,32	1,66	2,1833	33,3	1,541	1,495	29,7	73,9
34	74,16	19,36	5,98	0,35	1,67	2,2066	32,8	1,547	1,500	31,8	73,7
35	72,76	19,77	7,47	0,375	1,69	2,2590	32,0	1,550	1,500	25,8	73,2
36	71,54	19,60	8,86	0,39	1,70	2,2693	32,0	1,546	1,495	17,1	73,1
37	70,20	19,58	10,22	0,425	1,71	2,3400	31,3	1,546	1,495	—	72,3
38	68,82	19,56	11,63	0,45	1,73	2,3668	30,5	1,533	1,485	14,5	72,0
39	67,46	19,60	12,94	0,48	1,74	2,3693	31,0	1,555	1,505	11,5	72,0
40	65,98	19,95	14,00	0,52	1,76	2,3856	30,5	1,557	1,500	14,1	71,8

TABLE 14

Glasses with the Composition $B_2O_3 - Li_2O - BeO - MgO$ [63]

Glass No.	Composition, mol. %				$\frac{Li_2O+MeO}{B_2O_3}$	$\frac{O}{B}$	D of glass g/cm³	$V_{B_2O_3}$, cm³/mole of glass	N_D		Solubility in water, %	X-ray transparency, % (t = 1.0 cm)
	B_2O_3	Li_2O	BeO	MgO					of glass	B_2O_3		
30	67,4	5,0	5,8	21,8	0,48	1,74	2,26	33,6	1,550	1,515	4,8	74,3
31	82,0	1,4	6,2	1,5	0,22	1,61	2,09	34,1	1,507	1,475	96,5	77,1
38	72,3	10,0	5,9	11,8	0,38	1,69	2,10	34,8	1,528	1,485	9,7	75,8
46	77,2	15,2	6,1	1,5	0,3	1,65	2,21	32,5	1,522	1,485	86,0	77,2
53	68,0	14,5	5,8	11,7	0,47	1,73	2,14	34,5	1,537	1,485	10,8	75,7
60	60,0	13,9	5,6	20,5	0,67	1,83	2,19	36,5	1,551	1,490	3,8	—
61	72,9	19,7	5,9	1,5	0,37	1,69	2,15	34,0	1,525	1,475	66,0	77,2
68	64,3	18,9	5,6	11,2	0,56	1,78	2,28	32,5	1,552	1,490	5,2	75,7
75	56,4	18,1	5,4	20,1	0,77	1,89	2,13	36,2	1,560	1,490	4,2	74,3
76	69,0	23,9	5,7	1,4	0,45	1,73	2,19	33,2	1,540	1,480	63,5	77,2
83	60,7	22,9	5,5	10,9	0,64	1,83	2,24	33,3	1,560	1,495	6,1	75,7
90	53,0	22,1	5,3	19,6	0,89	1,94	—	—	1,541	1,450	9,0	—
91	65,0	28,0	5,6	1,4	0,54	1,77	2,04	36,0	1,554	1,490	41,2	77,0
98	57,0	26,8	5,4	10,8	0,75	1,87	2,33	32,0	1,560	1,485	8,3	75,5
120	63,0	5,3	11,0	20,7	0,59	1,79	2,23	34,0	1,540	1,490	2,8	74,0
128	67,6	9,5	11,5	11,4	0,48	1,74	2,14	34,7	1,533	1,480	5,5	75,5
135	59,3	9,2	11,0	20,5	0,67	1,83	2,34	32,7	1,555	1,500	2,3	74,0
136	72,2	14,6	11,7	1,5	0,39	1,69	2,15	33,6	1,523	1,480	49,2	77,0
143	63,8	14,0	11,2	11,0	0,57	1,79	2,19	34,0	1,516	1,450	42,0	75,4
150	55,8	13,5	10,7	20,0	0,79	1,89	2,42	31,0	1,563	1,520	1,5	74,0
151	68,3	19,0	11,3	1,4	0,47	1,73	2,01	36,7	1,531	1,470	44,3	76,8
158	60,1	18,2	10,9	10,8	0,66	1,83	2,22	34,0	1,552	1,410	5,6	75,4
165	52,6	17,6	10,3	19,5	0,90	1,95	2,15	36,6	1,566	1,500	2,0	73,9
166	64,6	23,0	11,0	1,4	0,55	1,77	2,15	34,0	1,538	1,490	39,1	76,7
173	56,7	22,2	10,6	10,5	0,77	1,89	2,17	35,0	1,555	1,480	5,5	75,4
181	60,8	27,0	10,7	1,5	0,64	1,82	2,15	31,5	1,545	1,500	31,0	76,8
188	53,5	26,0	10,3	10,2	0,87	1,93	2,24	34,0	1,560	1,480	6,1	75,3
210	59,0	4,5	16,2	20,3	0,70	1,85	2,11	36,6	1,547	1,500	1,7	73,8
218	63,2	9,2	16,6	11,0	0,58	1,79	2,04	37,0	1,537	1,500	30,0	75,2
225	55,4	8,9	15,9	19,8	0,80	1,90	2,24	34,3	1,550	1,490	2,5	73,7

Glass No.	Composition, mol. % B₂O₃	Li₂O	BeO	MgO	$\dfrac{Li_2O+MeO}{B_2O_3}$	$\dfrac{O}{B}$	D of glass g/cm³	$V_{B_2O_3}$, cm³/mole of glass	N_D of glass	N_D B₂O₃	Solubility in water, %	X-ray transparency, % (t = 1.0 cm)
226	67,7	14,1	16,8	1,4	0,48	1,74	2,20	32,0	1,558	1,520	1,8	76,6
233	60,0	13,5	16,0	10,5	0,67	1,84	2,26	33,2	1,543	1,480	3,6	75,1
240	52,1	13,0	15,6	19,3	0,93	1,96	2,27	34,5	1,563	1,520	0,7	73,7
241	64,0	18,3	16,4	1,3	0,82	1,78	2,10	35,2	1,538	1,480	12,6	76,5
248	56,2	17,6	15,8	10,4	0,78	1,88	2,11	36,3	1,552	1,480	2,4	75,1
255	49,0	17,0	15,2	18,8	1,02	2,01	2,48	31,0	1,572	1,510	2,5	73,7
256	60,4	22,3	16,0	1,3	0,66	1,83	2,23	34,8	1,543	1,475	13,8	74,8
263	53,0	21,4	15,4	10,2	0,89	1,94	2,36	32,0	1,558	1,485	4,3	75,1
271	57,0	26,3	15,4	1,3	0,76	1,82	2,10	35,8	1,549	1,470	18,5	76,4
278	50,0	25,0	15,0	10,0	1,00	2,00	2,36	32,2	1,572	1,490	3,2	75,0
308	59,0	9,0	21,4	10,6	0,70	1,84	2,16	35,2	1,534	1,475	1,7	74,9
315	52,0	8,5	20,5	19,0	0,93	1,96	2,37	32,5	1,559	1,500	0,5	74,0
323	55,8	13,1	20,8	10,3	0,79	1,90	2,32	32,5	1,545	1,480	1,7	74,8
330	48,6	12,6	20,1	18,7	1,05	2,01	2,16	36,7	1,560	1,490	1,4	73,8
331	60,0	17,7	21,0	1,3	0,67	1,83	2,14	34,9	1,540	1,475	10,5	76,1
338	52,6	17,0	20,4	10,0	0,90	1,92	2,24	34,3	1,552	1,480	1,1	74,8
346	56,6	21,5	20,6	1,3	0,77	1,89	2,12	35,6	1,546	1,485	7,3	76,1
353	49,6	20,7	19,9	9,8	1,01	2,00	2,26	34,4	1,563	1,485	1,4	74,8
364	53,4	25,2	20,1	1,3	0,88	1,94	2,15	35,3	1,549	1,460	11,4	76,2
368	46,6	24,4	19,4	9,6	1,14	2,06	2,29	34,0	1,556	1,480	1,0	74,8
405	48,3	8,3	24,9	18,5	1,07	2,04	2,32	—	1,550	1,490	4,6	73,1
420	45,3	12,3	24,3	18,1	1,21	2,10	2,36	33,4	1,563	1,485	0,21	73,1
428	59,1	16,5	24,6	9,8	1,03	2,01	2,21	36,3	1,559	1,480	0,7	74,5
435	42,7	15,9	23,7	17,7	1,34	2,17	2,30	31,9	1,563	—	0,9	73,1
436	53,0	20,8	25,0	1,2	0,89	1,94	2,25	—	1,553	—	—	—
443	46,3	20,1	24,1	9,5	1,16	2,08	2,23	35,2	1,566	1,485	0,8	74,5
450	40,2	19,4	23,2	17,2	1,49	2,23	2,31	35,0	1,577	1,490	1,2	73,1
451	50,0	24,5	24,3	1,2	1,00	2,00	2,24	34,0	1,549	1,460	6,2	75,9
158	43,5	23,6	23,5	9,4	1,30	2,15	2,19	36,1	1,568	1,455	1,0	74,5
480	47,8	4,2	29,6	18,4	1,09	2,07	2,33	33,6	1,550	1,485	0,61	73,0
548	40,6	23,0	27,4	9,0	1,46	2,23	2,28	35,0	1,566	1,465	0,81	74,1

TABLE 15

Glasses with the Composition $B_2O_3 - Li_2O - CaO$

Glass No.	Composition, mol.%			$\dfrac{Li_2O+CaO}{B_2O_3}$	$\dfrac{O}{B}$	D of glass g/cm³	$V_{B_2O_3}$ cm³/mole	N_D		$\alpha \cdot 10^{-7}_{20-320}$		Solubility in water, %
	B_2O_3	Li_2O	CaO					of glass	B_2O_3	of glass	B_2O_3	
1	83,6	10,7	5,7	0,20	1,60	1,90	38,4	1,515	1,480	58,2	26,2	гель
2	78,0	10,7	11,3	0,28	1,64	2,14	34,8	1,539	1,480	60,6	21,8	14,0
3	74,3	20,3	5,4	0,35	1,67	2,20	33,2	1,541	1,485	72,0	13,5	14,0
4	72,5	10,5	17,0	0,38	1,69	2,26	33,5	1,560	1,485	69,5	26,2	6,5
5	69,0	20,2	10,8	0,45	1,73	2,26	33,2	1,558	1,490	74,0	7,8	7,0
6	65,1	28,8	5,1	0,51	1,76	2,24	32,8	1,559	1,485	85,6	2,1	36,0
7	67,5	10,4	22,1	0,49	1,74	—	—	1,567	1,490	71,8	22,1	5,3
8	64,2	20,0	15,8	0,55	1,78	2,29	33,5	1,570	1,490	83,0	13,1	4,5
9	61,5	28,5	10,0	0,63	1,81	2,29	32,9	1,571	1,485	94,0	6,5	4,0
10	58,8	36,4	4,8	0,70	1,91	2,25	33,1	1,559	1,475	106,0	4,3	7,0
61	60,0	5,0	35,0	0,65	1,83	2,47	33,1	1,592	1,500	—	—	2,0
62	60,0	15,0	25,0	0,66	1,83	2,42	32,6	1,599	1,520	—	—	1,5
63	70,0	5,0	25,0	0,43	1,71	2,43	33,6	1,572	1,515	—	—	5,0
64	80,0	5,0	15,0	0,25	1,63	2,23	33,3	1,540	1,490	—	—	10,0

TABLE 16

Glasses with the Composition B_2O_3 – Li_2O – ZnO

Glass No.	Composition, mol. %			$\dfrac{Li_2O + ZnO}{B_2O_3}$	$\dfrac{O}{B}$	D of glass, g/cm³	$V_{B_2O_3}$, cm³/mole	N_D		$\alpha \cdot 10^{-7}_{20-320}$		Solubility in water, %
	B_2O_3	Li_2O	ZnO					of glass	B_2O_3	of glass	B_2O_3	
11	85,0	11,0	4,0	0,18	1,59	2,14	34,9	1,507	1,475	61,0	34,5	52,0
12	80,6	11,2	8,2	0,24	1,62	2,33	32,2	1,523	1,480	65,0	38,0	43,0
13	75,5	20,7	3,8	0,32	1,66	2,23	33,1	1,531	1,480	73,9	21,2	51,0
14	76,8	11,0	12,2	0,30	1,65	2,36	32,9	1,535	1,485	69,6	44,0	34,5
15	71,7	20,6	7,7	0,40	1,70	2,385	31,7	1,547	1,485	77,0	25,3	42,0
16	67,0	29,4	3,6	0,49	1,74	2,33	31,6	1,548	1,475	82,0	1,05	43,0
17	72,4	11,2	16,4	0,23	1,69	2,41	33,5	1,549	1,485	71,2	45,0	24,0
18	67,4	21,0	11,6	0,48	1,74	2,42	32,5	1,558	1,485	76,0	20,0	32,0
19	63,3	29,4	7,3	0,58	1,79	2,40	31,0	1,561	1,480	85,0	2,9	38,0
20	59,6	37,0	3,4	0,68	1,83	2,36	31,6	1,563	1,480	106,0	7,2	40,0
65	60,0	5,0	35,0	0,67	1,83	2,80	33,8	1,596	1,520	—	—	1,0
66	60,0	15,0	25,0	0,58	1,84	2,57	35,5	1,586	1,510	—	—	1,0
67	60,0	25,0	15,0	0,74	1,87	2,48	31,6	1,579	1,498	—	—	14,0
68	70,0	5,0	25,0	0,43	1,71	2,46	35,1	1,576	1,520	—	—	3,0
69	80,0	5,0	15,0	0,32	1,63	2,40	33,8	1,541	1,510	—	—	15,0

TABLE 17

Glasses with the Composition B_2O_3 — Li_2O — SrO

| Glass No. | Composition, mol. % | | | $\dfrac{Li_2O + SrO}{B_2O_3}$ | $\dfrac{O}{B}$ | D of glass g/cm³ | $V_{B_2O_3}$ cm³/mole | N_D | | $\alpha \cdot 10^{-7}_{20-320}$ | | Solubility in water, % |
	B_2O_3	Li_2O	SrO					of glass	B_2O_3	of glass	B_2O_3	
21	85,8	11,0	3,2	0,17	1,58	2,10	34,9	1,508	1,475	65,0	35,0	69,0
22	82,4	11,2	6,4	0,21	1,61	2,21	34,0	1,523	1,480	70,0	36,5	38,2
23	76,9	21,0	3,0	0,31	1,65	2,16	34,2	1,538	1,485	82,0	27,2	45,6
24	78,6	11,4	10,0	0,27	1,63	2,44	31,7	1,553	1,500	65,0	24,5	11,6
25	72,6	21,2	6,2	0,375	1,69	2,52	30,6	1,540	1,480	80,0	18,5	14,5
26	67,8	29,4	2,8	0,48	1,73	2,44	30,0	1,554	1,485	90,0	9,4	17,8
27	75,0	11,6	13,4	0,33	1,67	2,55	31,6	1,552	1,490	70,0	25,0	9,4
28	69,0	21,6	9,4	0,44	1,72	2,63	29,5	1,562	1,490	80,0	11,0	9,5
29	64,5	29,8	5,7	0,55	1,77	2,59	29,1	1,564	1,485	87,5	—2,4	9,5
30	60,0	37,3	2,7	0,66	1,82	2,72	26,5	1,561	1,485	101,0	—6,7	13,5
70	60,0	5,0	35,0	0,66	1,83	3,10	31,4	1,596	1,485	—	—	3,0
71	60,0	15,0	25,0	0,66	1,83	3,06	29,0	1,594	1,495	—	—	4,0
72	60,0	25,0	15,0	0,67	1,83	2,84	29,0	1,587	1,490	—	—	6,8
73	67,5	12,5	20,0	0,30	1,74	2,91	29,0	1,585	1,510	—	—	5,6
74	70,0	5,0	25,0	0,44	1,71	3,03	28,5	1,589	1,505	—	—	8,0
75	80,0	5,0	15,0	0,25	1,63	2,80	28,5	1,541	1,485	—	—	2,0

TABLE 18

Glasses with the Composition $B_2O_3 - Li_2O - CdO$

Glass No.	Composition, mol, %			$\dfrac{Li_2O+CdO}{B_2O_3}$	$\dfrac{O}{B}$	D of glass g/cm³	$V_{B_2O_3}$ cm³/mole	N_D		$\alpha \cdot 10^{-7}_{20-320}$		Solubility in water, %
	B_2O_3	Li_2O	CdO					of glass	B_2O_3	of glass	B_2O_3	
31	86,2	11,2	2,6	0,16	1,58	2,16	34,0	1,513	1,480	56,0	26,4	78,0
32	83,3	11,4	5,3	0,2	1,6	2,26	33,6	1,532	1,490	57,0	25,0	17,5
33	76,5	21,0	2,5	0,31	1,65	2,28	32,4	1,538	1,480	71,0	15,0	28,0
34	80,3	11,6	8,1	0,25	1,62	2,49	31,5	1,540	1,485	60,0	24,0	12,5
35	73,7	21,3	5,0	0,36	1,68	2,47	30,9	1,555	1,490	67,0	5,1	13,0
36	68,0	29,7	2,3	0,47	1,73	2,47	29,8	1,557	1,485	74,0	−12,2	26,5
37	77,0	11,9	11,1	0,3	1,65	2,66	30,7	1,570	1,495	61,0	21,0	3,7
38	70,5	21,9	7,6	0,42	1,71	2,80	27,8	1,575	1,500	71,0	4,3	8,5
39	65,0	30,8	4,8	0,54	1,77	2,97	25,0	1,570	1,490	76,0	−17,0	10,5
40	60,4	37,4	2,2	0,66	1,83	2,99	23,7	1,573	1,495	85,0	−30,0	28,0
76	60,0	5,0	35,0	0,67	1,83	3,60	30,0	1,653	1,520	—	—	0,3
77	60,0	15,0	25,0	0,67	1,83	3,30	29,1	1,630	1,510	—	—	0,5
78	60,0	25,0	15,0	0,67	1,83	3,15	27,5	1,596	1,480	—	—	2,0
79	67,5	12,5	20,0	0,49	1,74	3,12	29,5	1,597	1,500	—	—	0,85
80	70,0	5,0	25,0	0,42	1,71	3,22	29,6	1,604	1,510	—	—	0,5
81	80,0	5,0	15,0	0,25	1,60	2,785	30,4	1,559	1,490	—	—	2,0

TABLE 19

Glasses with the Composition B_2O_3 – Li_2O – BaO

Glass No.	Composition, mol. %			$\frac{Li_2O+BaO}{B_2O_3}$	$\frac{O}{B}$	D of glass g/cm³	$V_{B_2O_3}$ cm³/mole	N_D		$\alpha \cdot 10^{-7}_{20-320}$		Solubility in water, %
	B_2O_3	Li_2O	BaO					of glass	B_2O_3	of glass	B_2O_3	
41	86,8	11,1	2,1	0,16	1,58	2,45	29,6	1,503	1,480	68,0	39,0	87,0
42	84,1	11,5	4,4	0,19	1,59	2,46	30,7	1,518	1,475	57,0	20,5	75,0
43	77,0	21,0	2,0	0,30	1,65	2,49	29,3	1,530	1,480	73,0	16,0	60,0
44	81,2	11,8	7,0	0,23	1,61	2,49	31,4	1,528	1,475	69,0	28,5	61,5
45	74,2	21,6	4,2	0,35	1,67	2,52	30,1	1,546	1,485	73,0	8,2	28,5
46	68,2	29,8	2,0	0,46	1,72	2,54	28,9	1,543	1,470	83,0	—1,3	19,5
47	78,5	12,1	9,4	0,27	1,64	2,52	32,4	1,546	1,485	70,0	23,5	24,0
48	71,4	22,2	6,4	0,40	1,70	2,61	30,2	1,550	1,480	73,0	0,3	27,5
49	65,7	30,3	4,0	0,53	1,76	2,62	28,0	1,560	1,480	83,0	—10,2	18,8
50	60,6	37,6	1,8	0,65	1,83	2,93	24,1	1,557	1,450	95,0	—16,7	27,0
82	60,0	5,0	35,0	0,67	1,84	3,91	27,6	1,615	1,460	—	—	5,0
83	60,0	15,0	25,0	0,67	1,83	3,73	25,9	1,608	1,475	—	—	8,0
84	60,0	25,0	15,0	0,67	1,83	3,34	26,0	1,596	1,485	—	—	7,8
85	70,0	5,0	25,0	0,48	1,71	3,40	28,6	1,579	1,465	—	—	13,5
86	70,0	15,0	15,0	0,43	1,71	3,14	27,5	1,576	1,485	—	—	9,5
87	80,0	5,0	15,0	0,25	1,63	2,88	30,0	1,552	1,480	—	—	23,0

144

TABLE 20

Glasses with the Composition B_2O_3 – Li_2O – PbO

Glass No.	Composition, mol. %			$\frac{Li_2O+PbO}{B_2O_3}$	$\frac{O}{B}$	D of glass g/cm³	$V_{B_2O_3}$ cm³/mole	N_D		$\alpha \cdot 10^{-7}_{20-320}$		Solubility in water, %
	B_2O_3	Li_2O	PbO					of glass	B_2O_3	of glass	B_2O_3	
51	87,3	11,2	1,5	0,15	1,57	2,12	34,8	1,515	1,480	62,0	34,0	87,0
52	85,2	11,7	3,1	0,17	1,58	2,16	35,7	1,525	1,475	70,0	40,0	64,0
53	77,4	21,2	1,4	0,29	1,65	2,26	33,8	1,537	1,480	70,0	27,0	78,0
54	83,2	12,0	4,8	0,20	1,60	2,26	35,6	1,536	1,480	73,0	41,2	40,0
55	75,0	22,0	3,0	0,33	1,66	2,28	33,9	1,555	1,485	83,0	26,3	38,0
56	69,0	29,7	1,3	0,46	1,72	2,28	32,9	1,555	1,480	88,0	8,8	—
57	80,8	12,5	6,7	0,235	1,61	2,31	36,4	1,555	1,480	76,0	41,5	39,0
58	73,0	22,5	4,5	0,37	1,68	2,41	33,5	1,563	1,485	85,0	41,2	36,0
59	66,5	30,7	2,8	0,50	1,75	2,52	30,7	1,563	1,480	95,0	12,5	48,0
60	61,0	37,7	1,3	0,64	1,82	2,55	32,0	1,567	1,475	105,0	2,1	53,0
88	60,0	5,0	35,0	0,67	1,83	4,33	31,9	1,715	1,460	—	—	1,5
89	60,0	15,0	25,0	0,67	1,83	3,99	31,0	1,672	1,470	—	—	1,6
90	60,0	25,0	15,0	0,67	1,83	3,68	27,2	1,674	1,495	—	—	12,0
91	70,0	5,0	25,0	0,43	1,71	3,71	31,6	1,666	1,490	—	—	1,8
92	70,0	15,0	15,0	0,43	1,71	—	—	1,638	1,520	—	—	4,0
93	80,0	5,0	15,0	0,25	1,63	2,96	33,3	1,606	1,500	—	—	7,0
94	90,0	5,0	5,0	0,11	1,56	2,22	35,5	1,540	1,485	—	—	36,0

TABLE 21

Physicochemical Properties of Glasses Investigated in Relation to the Al_2O_3 and SiO_2 Added

Glass No.	Calculated composition, wt.%				Additives in parts by weight		X-ray transparency, %	Density, g/cm³	Refractive index	Solubility in water, %	Remarks
	B_2O_3	BeO	MgO	Li_2O	Al_2O_3	SiO_2					
9	80,0	2,5	15,0	2,5	—	—	74,36	2,26	1,550	4,80	Series I: additives in weight percent over 100% composition of starting glass
11	80,0	2,5	15,0	2,5	2,0	—	73,96	2,28	1,549	2,63	
12	80,0	2,5	15,0	2,5	5,0	—	73,44	2,43	1,545	3,55	
15	80,0	2,9	11,5	5,6	2,5	10,0	74,18	2,28	1,540	4,00	
17	77,5	5,0	15,0	2,5	—	—	74,05	2,23	1,540	2,77	
19	77,5	2,5	15,0	5,0	5,0	—	73,40	2,13	1,545	3,03	
21	75,0	7,5	15,0	2,5	—	—	73,77	2,10	1,547	1,68	
22	75,0	2,5	15,0	7,5	—	—	74,20	2,19	1,551	3,74	
23	75,0	2,5	15,0	7,5	5,0	—	73,34	2,24	1,552	2,87	
27	70,0	10,0	15,0	5,0	—	—	74,02	2,37	1,558	0,50	
28	70,0	10,0	15,0	5,0	2,0	—	73,09	2,34	1,555	1,04	
29	70,0	10,0	15,0	5,0	5,0	—	72,59	2,38	1,556	0,78	
31	69,0	15,0	1,0	15,0	2,0	—	75,19	2,23	1,553	5,21	
32	67,5	15,0	15,0	2,5	—	—	72,80	2,33	1,550	0,61	
33	67,5	15,0	15,0	2,5	10,0	—	72,05	2,23	1,562	0,90	
34	67,5	12,5	15,0	5,0	—	—	73,16	2,31	1,550	4,52	
35	65,0	12,5	15,0	7,5	—	—	73,13	2,36	1,563	0,21	

TABLE 21 (continued)

Glass No.	Calculated composition, wt. %				Additives in parts by weight		X-ray transparency, %	Density, g/cm³	Refractive index	Solubility in water, %	Remarks
	B₂O₃	BeO	MgO	Li₂O	Al₂O₃	SiO₂					
36	65,0	12,5	15,0	7,5	2,0	—	72,80	2,19	1,565	1,50	Series I: additives in weight percent over 100% composition of starting glass
37	65,0	12,5	15,0	7,5	5,0	—	72,30	2,36	1,560	0,47	
38	64,5	12,5	8,0	15,0	—	—	74,46	2,19	1,568	1,00	
39	64,5	12,5	8,0	15,0	2,0	—	74,09	2,40	1,563	2,62	
40	64,5	12,5	8,0	15,0	5,0	—	73,55	2,16	1,565	2,87	
41	64,5	12,5	8,0	15,0	10,0	—	72,54	2,23	1,557	2,80	
43	62,0	15,0	8,0	15,0	—	—	74,15	2,28	1,566	0,81	
44	62,0	15,0	8,0	15,0	2,0	—	73,85	2,37	1,572	1,25	
45	62,0	15,0	8,0	15,0	10,0	—	71,58	2,20	1,563	3,00	
49	60,0	15,0	15,0	10,0	10,0	—	71,10	2,33	1,577	0,32	
57	57,5	12,5	15,0	15,0	5,0	—	72,17	2,32	1,568	0,32	
62/9	72,5	—	10,0	2,5	5,0	10,0	74,25	2,14	1,535	15,0	Series II: additives as replacements. Denominator – No. of original glass
63/9	75,0	2,5	5,0	2,5	5,0	10,0	74,81	2,09	1,508	16,30	
26	72,5	7,5	15,0	5,0	—	—	73,74	2,24	1,550	2,52	
66 26	67,5	2,5	10,0	5,0	5,0	10,0	73,62	2,36	1,543	5,25	
67 27	70,0	5,0	10,0	5,0	—	10,0	73,96	2,36	1,543	2,00	
68 32	60,0	5,0	12,5	2,5	10,0	10,0	72,53	2,36	1,552	0,90	
70 35	65,0	5,0	10,0	7,5	2,5	10,0	73,95	2,17	1,553	1,76	

TABLE 22

Compositions of Glasses and Their Thermal Expansion with SiO₂ and Al₂O₃ Added

Glass No.	Chemical composition, wt. %				Additives, wt. %		$\alpha \cdot 10^{-7}$
	B_2O_3	Li_2O	BeO	MgO	Al_2O_3	SiO_2	
37	88,0	3,8	5,0	3,2	5,0	—	58
39	88,0	3,8	5,0	3,2	20,0	—	67
40	88,0	3,8	5,0	3,2	30,0	—	69
42	88,0	3,8	5,0	3,2	50,0	—	75
43	89,8	3,2	4,3	2,7	1,0	—	58
46	89,8	3,2	4,3	2,7	20,0	—	68
47	89,8	3,2	4,3	2,7	40,0	—	68
51	90,5	3,0	4,0	2,5	5,0	—	51
52	90,5	3,0	4,0	2,5	10,0	—	58
57	85,0	12,0	3,0	—	2,5	1,0	70
59	85,0	12,0	3,0	—	2,5	10,0	71
61	85,0	12,0	3,0	—	2,5	20,0	72
63	87,0	10,4	2,6	—	2,5	1,0	52
64	87,0	10,4	2,6	—	2,5	5,0	65
66	87,0	10,4	2,6	—	2,5	15,0	67
68	87,0	10,4	2,6	—	2,5	25,0	69
69	88,0	9,6	2,4	—	2,5	1,0	56
70	88,0	9,6	2,4	—	2,5	5,0	62
71	88,0	9,6	2,4	—	2,5	10,0	63
73	88,0	9,6	2,4	—	2,5	20,0	72
74	88,0	9,6	2,4	—	2,5	25,0	74
146	80,0	2,5	2,5	15,0	1,0	—	46
147	80,0	2,5	2,5	15,0	5,0	—	52
148	80,0	2,5	2,5	15,0	10,0	—	58
149	80,0	2,5	2,5	15,0	20,0	—	58

TABLE 23

Constants for Calculating the X-ray Transparency of Glasses (for the elements of the basic oxides — glass components)

Main indices		Elements						
		Li	Be	B	O	Mg	Al	Si
Atomic number (Z)		3	4	5	8	12	13	14
Atomic weight (A)		6,94	9,02	10,82	16,00	24,32	26,97	28,06
Oxides (M_mO_n)		Li_2O	BeO	B_2O_3	—	MgO	Al_2O_3	SiO_2
Weight fraction	M_m	0,4645	0,3605	0,3107	—	0,6032	0,5291	0,4672
	O_n	0,5355	0,6395	0,6893	—	0,3968	0,4709	0,5328
Constants for calculation (depending on Z and A)	C	0,1507	0,3689	0,6006	3,1610	11,65	14,69	19,19
	D	$493,60 \times 10^{-6}$	$2133,0 \times 10^{-6}$	0,006791	0,07739	0,5876	0,8583	1,286
$F = \dfrac{Z}{A}$		0,4322	0,4434	0,4621	0,5000	0,4934	0,4820	0,4989

TABLE 23 (continued)

Main indices		Elements						
		Li	Be	B	O	Mg	Al	Si
ω_1 for wavelengths in angstroms (A)	0,01 Å	0,0488	0,0501	0,0522	0,0565	0,05576	0,05446	0,0564
	0,05 Å	0,0977	0,1002	0,1045	0,1133	0,1129	0,1089	0,1151
	0,1 Å	0,1208	0,1241	0,1295	0,1427	0,1493	0,1491	0,1583
	0,5 Å	0,1767	0,2078	0,2433	0,5728	1,6004	1,9589	2,5005
	1,0 Å	0,3187	0,5384	0,7726	3,2775	11,1533	14,0170	18,097
	5,0 Å	18,8260	44,9558	71,0146	346,955	1097,1264	1300,0	1595,0

Mass absorption coefficient of each element $\omega = \dfrac{\mu}{\rho} \equiv C\lambda^3 - D\lambda^4 + EF,$

where λ is the wavelength in angstroms, E is the local scattering of electrons, F = Z/A, and ρ is the density.

At $\lambda = 0,01$ Å	E = 0,113	$\lambda^3 = 1,0 \times 10^{-6}$	$\lambda^4 = 1,0 \times 10^{-8}$
" 0,05 Å	E = 0,226	$12,5 \times 10^{-6}$	625×10^{-8}
" 0,1 Å	E = 0,279	0,001	0,0001
" 0,5 Å	E = 0,365	0,125	0,0625
" 1,0 Å	E = 0,387	1,0	1,0
" 5,0 Å	E = 0,398	125,0	625,0

The absorption coefficient of each element (or oxide) $\mu = \omega \rho$.

The mass absorption coefficient of each oxide $\omega M_m O_n = f_m \omega_m + f_0 \omega_0$

where ω is the mass absorption coefficient of the elements M and O and f_m and f_0 are the weight fractions of them.

TABLE 24

Constants for Calculating the X-ray Transparency of Glasses (for oxides)

Main indices		Oxides					
		Li_2O	BeO	B_2O_3	MgO	Al_2O_3	SiO_2
Density – ρ (156)		2,01	3,0	1,81	3,58	3,80	2,203
ω of oxides at λ =	0,01 Å	0,0528	0,0542	0,0541	0,0560	0,0564	0,0556
	0,05 Å	0,106	0,1085	0,1105	0,1130	0,1110	0,1141
	0,1 Å	0,1325	0,1359	0,1385	0,1466	0,1460	0,1499
	0,5 Å	0,389	0,4412	0,4704	1,1925	1,306	1,473
	1,0 Å	1,9031	2,29	2,499	8,033	8,952	10,203
	5,0 Å	194,56	238,0	261,0	675,5	851,0	930,0
μ of oxides at λ =	0,01 Å	0,1061	0,1626	0,0997	0,2	0,210	0,123
	0,05 Å	0,2131	0,3255	0,2001	0,405	0,422	0,251
	0,1 Å	0,267	0,408	0,251	0,525	0,555	0,330
	0,5 Å	0,782	1,3236	0,8154	4,269	4,963	3,240
	1,0 Å	3,825	6,87	4,52	28,76	34,0	22,40
	5,0 Å	391,0	714,0	472,0	2418,0	3234,0	2046,0

TABLE 24 (continued)

Main indices	Oxides					
	Li₂O	BeO	B₂O₃	MgO	Al₂O₃	SiO₂
Half thickness at λ =						
0,01 Å	6,53	4,20	6,95	3,47	3,30	5,63
0,05 Å	3,26	2,13	3,46	1,71	1,64	2,76
0,1 Å	2,60	1,70	2,76	1,32	1,25	2,10
0,5 Å	0,89	0,53	0,82	0,16	0,14	0,22
1,0 Å	0,20	0,100	0,16	0,025	0,02	0,03

$$d_{1/2} = \frac{\ln 2}{\mu} = \frac{0,693}{\mu} \ \text{cm}$$

$\mu_{glass} = \rho_{glass} \cdot \Sigma \left(f_{oxides} \cdot \omega_{oxides} \right).$

Footnote. We calculated the constants ω of the elements and ω and μ of the oxides for $\lambda = 0,01$ A, 0.05 A, 0.5 A, 1.0 A, and 5.0 A. The other data are taken from appropriate literature sources [69, 70, 156].

LITERATURE CITED

[1] M. A. Bezborodov, M. V. Lomonosov and His Work on the Chemistry and Technology of Silicates (Acad. Sci. USSR Press, 1948) [in Russian].

[2] D. I. Mendeleev, Glass Production (St. Petersburg, 1864) [in Russian].

[3] B. M. Kedrov, Uspekhi Khimi [Advances in Chemistry] XXI, No. 8, 1952.

[4] Glass Technology, edited by I. I. Kitaigorodskii (Building Materials Press, 1951) [in Russian].

[5] I. I. Kitaigorodskii, "Soviet glass manufacture and its most immediate problems," Uspekhi Khimi [Advances in Chemistry] XXIII, No. 4, 1954.

[6] The Structure of Glass, edited by M. A. Bezborodov (State Chem. Tech. Press, 1933) [in Russian].

[7] M. W. Trawers, Kolloid-Z. 28, 1921.

[8] E. Berger, Glastech. Ber. 5, 1927.

[9] F. Eckert, Z. tech. Physik. 7, 1926.

[10] W. Eitel, Physical Chemistry of Silicates (United Sci. Tech. Press, Chem. Theor., 1936) [Russian translation].

[11] A. F. Wells, The Structure of Inorganic Substances (Foreign Lit. Press, 1948) [Russian translation].

[12] A. A. Ginzberg, Experimental Petrography (Leningrad State University Press, 1951).

[13] The Structure of Glass (Acad. Sci. USSR Press, 1955) [in Russian].

[14] B. Ya. Blyumberg, Introduction to the Physical Chemistry of Glass (State Chem. Press 1940) [in Russian].

[15] R. Schmidt, Der Praktische Glas Schmelzer (Leipzig, 1953).

[16] S. Scholes, Modern Glass Practice (Chicago, 1935).

[17] G. Morey, The Properties of Glass (New York, 1938).

[18] C. Phillips, Glass, The Miracle Maker (New York, 1941).

[19] P. P. Kobeko, Amorphous Substances (Acad. Sci. USSR Press, 1952) [in Russian].

[20] O. K. Botvinkin, Physical Chemistry of Silicates (Building Materials Press, 1955) [in Russian].

[21] K. S. Evstrop'ev and N. A. Toropov, Physical Chemistry of Silicates (Building Materials Press, 1956) [in Russian].

[22] K. S. Evstrop'ev, "A history of the development of the hypothesis on the crystalline structure of glass, Steklo i Keram. [Glass and Ceramics] 5, 1954.

[23] V. P. Barzakovskii, Steklo i Keram. [Glass and Ceramics] 2, 1954.

[24] L. I. Demkina, Steklo i Keram. [Glass and Ceramics] 2, 1954.

[25] V. M. Gol'dshmidt, Crystallochemistry (United Sci. Tech. Press, 1937) [in Russian].

[26] G. B. Bokii, Introduction to Crystallochemistry (Moscow State University Press, 1954) [in Russian].

[27] I. V. Grebenshchikov and O. S. Molchanova, Zhur. Obshchei Khim. [J. Gen. Chem.] 12, 1942.

[28] O. S. Molchanova, Priroda [Nature] 4, 1947.

[29] E. A. Porai-Koshits , Zhur. Obshchei Khim. [J. Gen. Chem] 12, 1942; Doklady Akad. Nauk SSSR [Proc. Acad. Sci. USSR] 40, 1944.

[30] S. P. Zhdanov, Doklady Akad. Nauk SSSR [Proc. Acad. Sci. USSR] 82, 1952.

[31] E. A. Porai-Koshits, Uspekhi Khimi [Advances in Chemistry] 16, 1947.

[32] E. A. Porai-Koshits, Uspekhi Fiz. Nauk [Advances in Physical Science] 39, 1949.

[33] A. A. Appen, Uspekhi Khim. [Advances in Chemistry] XXI, No. 4, 1952.

[34] A. A. Appen, Steklo i Keram. [Glass and Ceramics] 3, 1954.

[35] A. A. Appen, Zhur. Priklad. Khim. [J. Appl. Chem.] XXVII, 2, 1954.

[36] B. Warren and A. Loring, J. Am. Ceram. Soc. 18, 1935.

[37] R. C. Evans, Introduction to Crystallochemistry (State Chem. Press, 1948) [in Russian].

[38] N. V. Belov, The Structure of Ionic Crystals and Metallic Phases (Acad. Sci. USSR Press, 1947) [in Russian].

[39] Chemistry of Borates [Acad. Sci. Latvian SSR Press, 1953) [in Russian].

[40] A. G. Betekhtin, Mineralogy (State Geol. Press, 1950) [in Russian].

[41] V. B. Nekrasov, General Chemistry (State Chem. Press, 1945) [in Russian].

[42] 'Chemists' Handbook I (State Chem. Press, 1951) [in Russian].

[43] W. Zachariasen, Z. Krist. 78, 289, 1931; 88, 150, 1934.

[44] W. Zachariasen, J. Chem. Phys. 5, 919, 1937.

[45] A. V. Nikolaev, Physicochemical Study of Natural Borates (Acad. Sci. USSR Press, 1947) [in Russian].

[46] V. Sobolev, "General problems of the structure of borates" Mineralogical Collection of L'vov Geological Society, No. 3, 1949 [in Russian].

[47] K. Fajans and S. Barber, J. Am. Chem. Soc. 74, 2761, 1952.

[48] V. V. Tarasov and E. F. Stroganov, Trans. D. I. Mendeleev Chemicotechnological Institute, Moscow, No. XXI (Building Materials Press, 1956).

[49] A. G. Kurnakova, Izvest. Akad. Nauk SSSR, Ser. Fiz-Khim. [Bull. Acad. Sci. USSR, Phys. Chem. Ser] XV, 1947.

[50] A. G. Kurnakova and A. V. Nikolaev, Izvest. Akad. Nauk SSSR, Otdel. Khim. Nauk [Bull. Acad. Sci. USSR, Div. Chem. Sci.] 4, 1948.

[51] A. G. Kurnakova, Izvest. Akad. Nauk SSSR, Ser. Fiz-Khim. [Bull. Acad. Sci. USSR, Phys. Chem. Ser.] XVIII, 1949.

[52] D. S. Belyankin, N. A. Toropov and V. V. Lapin, Physicochemical Systems of Silicate Technology (Building Materials Press, 1954) [in Russian].

[53] N. A. Toropov and P. F. Konovalov, Zhur. Fiz. Khim. [J. Phys. Chem] 4, 1940.

[54] H. M. Davis and M. A. Knight, J. Am. Ceram. Soc. 28 (4), 100, 1945.

[55] Phase Diagram for Ceramists, J. Am. Ceram. Soc., Nov. 1, pt. II, 1947.

[56] C. Mazetti and F. de Carli, Gazz. chim. ital. 56, 23, 1926.

[57] E. Korder, Z. anorg. Chem. 241 [1], 1939.

[58] R. I. Bresker, Investigation of Optical and Dielectric Properties of Boron Glasses, Author's abstract of dissertation (State Optical Institute, 1949).

[59] R. I. Bresker and K. S. Evstrop'ev, Zhur. Priklad. Khim. [J. Appl. Chem.] XXV, No. 9, 1952.

[60] L. Ya. Mazelev, Study of the $B_2O_3 - BeO - Li_2O$ System in the Glassy State. Dissertation, 1939.

[61] L. Ya. Mazelev, Zhur. Priklad. Khim. [J. Appl. Chem.] XIII, No. 9, 1940.

[62] L. Ya. Mazelev, Zhur. Priklad. Khim. [J. Appl. Chem.] 3, 1953.

[63] L. Ya. Mazelev, Izvest. Akad. Nauk Belorus. SSR [Bull. Acad. Sci. Byelorussian SSR] 3, 1952.

[64] L. Ya. Mazelev, Trudy Belorus. Polytekh. Inst. [Trans. Byelorussian Polytechnic Institute] No. V, 1953.

[65] L. Ya. Mazelev, Izvest. Akad. Nauk Belorus. SSSR [Bull. Acad. Sci. Byelorussian SSR] 4, 1953.

[66] L. Ya. Mazelev, Trans. Byelorussian Polytechnic Institute, No. 55, 1956.

[67] L. Ya. Mazelev, Borolithium Glasses (Processes and Reactions of Glass Formation, Structure and Properties (Byelorussian State University Press, 1957) [in Russian].

[68] L. Ya. Mazelev, Composition — Property Diagrams of Borolithium Glasses , Trudy NIISM Belorus. SSR [Trans. Sci. Res. Inst. of Glass Byelorussian SSR] 6, 1957.

[69] "Glasses transparent to x-rays," Glass Ind. 26, 8, 1945.

[70] N. J. Kreidl, Glass Ind. 31, 3, 1950.

[71] Kuan Han-Sun, Glass Ind. 24, 1943.

[72] "Beryllium in glass," J. Soc. Glass Technol. 123, 1943.

[73] B. Laurent, Verres et refractoirs, 7, 3, 1953.

[74] S. English, J. Soc. Glass Technol. 8 (31) 1924.

[75] H. Menzel and S. Slivinski, Z. anorg. u. allgem. Chemie, 249 (4), 1942.

[76] W. Ziegler and M. Wellmann, Z. tech. Physik XIV, 7, 1933.

[77] H. Menzel and J. Adam, Glastrch. Ber. 22 (12), 1949.

[78] C. and F. Lindemann, Z. für Röntgenkunde, 13, 141, 1911.

[79] O. K. Botvinkin, Steklo i Keram. [Ceramics and Glass] 4, 1931.

[80] S. D. Gertsriken and K. A. Tanchakivskii, Author's Certificate No. 47050, class 32 B, 2.

[81] S. D. Gertsriken and M. A. Revutskaya, J. Phys. Chem. IV, No. 7, 1936.

[82] H. R. Moulton, Ceram. Abstr. 22 (1) 1943.

[83] K. H. Sun and T. E. Callear, Ceram. Abstr. Febr., 1948.

[84] W. H. Armistead, Ceram. Abstr. May, 1948.

[85] K. H. Sun and L. L. Sun, Glass Ind. 31, 10, 1950.

[86] L. M. Melnick, H. M. Safford, K. H. Sun and A. Silverman, J. Am. Ceram. Soc. 34, No. 3, 1951.

[87] G. F. Brewster and N. J. Kreidl, J. Am. Ceram. Soc. 35, 10, 1952.

[88] P. Debay, J. Appl. Phys. 20, 1949.

[89] Ceramic Industry 2, 1939.

[90] L. Biscoe and B. Warren, J. Am. Ceram. Soc. 24, 287, 1938.

[91] L. Shartsis, et al., J. Amer. Ceram. Soc. 36, 10, 1953; Khim. i. Khim. Tekhn. [Chemistry and Chemical Technology] 6, 1954.

[92] Tosio Abe, J. Am. Ceram. Soc. 35, 11, 1952.

[93] J. T. Littlton, J. Am. Ceram. Soc. 10, 4, 1927.

[94] B. W. King and A. J. Andrews, J. Am. Ceram. Soc. 23, 8, 1940.

[95] G. Gehlhoff and M. Thomas, Ceram. Abstr. 6 (12), 1927.

[96] J. Jamamoto and T. Kishii, Ceram. Abstr. March, 1952.

[97] B. E. Warren, J. Am. Ceram. Soc. 22, 180, 1939.

[98] B. E. Warren, J. Am. Ceram. Soc. 24, 256, 1941.

[99] E. J. Gooding and W. E. S. Turner, J. Soc. Glass Technol. 18, 69, 1934.

[100] W. E. S. Turner and F. Winks, J. Soc. Glass Technol. 14, 53, 1930.

[101] G. Morey and H. E. Merwin, J. Am. Ceram. Soc. 58 (11) 1936.

[102] Dralle-Keppeler. Glass Production, Vol. I, pt. 1 (Prodosilikata Press, 1928) [in Russian].

[103] V. L. Indenbom, Doklady Akad. Nauk SSSR [Proc. Acad. Sci. USSR] 89, 3, 1953.

[104] V. I. Rakov, Electronic x-ray Tubes (State Power Press, 1952) [in Russian].

[105] A. A. Appen, Zhur. Priklad. Khim. [J. Appl. Chem.] XXVI, 6, 1953.

[106] E. A. Porai-Koshits, Doklady Akad. Nauk SSSR [Proc. Acad. Sci. USSR] 36, 1942.

[107] K. Grjotheim and J. Krogh-Moe, Norske Widenskab. Selskabs Skrifter 27, 1954; 18 (Trondheim, 1955).

[108] B. I. Kogan, Khim. Nauk i Prom. [Chemical Sciences and Industry] 5, 1, 1956.

[109] B. A. Sakharov, Khim. Nauk i Prom. [Chemical Sciences and Industry] 5, 1, 1956.

[110] "Lithium," Ceram. Ind. 2, 1933.

[111] L. Navias, J. Am. Ceram. Soc. 18, 17, 1935.

[112] H. E. Simpson, Glass Ind. 9, 1955.

[113] F. I. Shamrai, Lithium and Its Alloys (Acad. Sci. USSR Press, 1952) [in Russian].

[114] P. P. Budnikov and A. M. Cherepanov, Uspekhi Khimi. [Advances in Chemistry].

[115] N. V. Solomin, Zhur. Teoret. Fiz. [J. Theo. Phys.] VIII, No. 6, 1938.

[116] A. Winkelmann and O. Schott, Ann. der Physik u. Chemie 51, 1894.

[117] S. C. Watterton and W. E. S. Turner, J. Soc. Glass Technol. 18, 1935.

[118] G. Heyne, Angew. Chemie 28, 1933.

[119] M. B. Reifman, Khim. Nauk i Prom. [Chemical Science and Industry] 5, 1, 1956.

[120] E. Zschimmer, Theorie der Glasschmelzkunst, I (Jena, 1923).

[121] A. Karsten, Verre et silicates inds. 7, 1936.

[122] Chi Fang Lai and A. Silvermann, J. Am. Ceram. Soc. 11, 1929.

[123] Chi Fang Lai and A. Silvermann, J. Am. Ceram. Soc. 13, 1930.

[124] A. Mauri, J. Soc. Glass Technol. XXI, 83, Abstr. 46, 1937.

[125] K. A. Becker, Sprechsaal, 10, 11, 12, 13, 14, 15, 16, 17 (1930).

[126] C. Gottfried, Glastech. Ber. 8, 1930.

[127] "Beryl" Ceram. Ind. 2, 1939.

[128] F. Eisenloeffel, German pat. 444749, Kl. 32, b., gr. I.

[129] A. V. Bleininger and F. H. Riddle, J. Am. Ceram. Soc. 1919, p. 564.

[130] L. Ainsworth, J. Soc. Glass Technol. XXXVIII, 1954.

[131] "Boric Oxide" Ceram. Ind. 2, 1939.

[132] L. G. Berg, Rapid Quantitative Phase Analysis (Acad. Sci. USSR Press, 1952) [in Russian].

[133] W. F. Hillebrand and G. E. Lundell, Practical Handbook of Inorganic Analysis (United Sci. Tech. Press, 1937) [Russian translation].

[134] I. M. Kolthoff and E. B. Sandell, Quantitative Analysis (State Sci. Tech. Press, 1948) [Russian translation].

[135] Berl'-Lunge, Chemicotechnical Methods of Analysis (United Sci. Tech. Press, NKTP USSR, 1938) [in Russian].

[136] A. Bender, Preparation and Testing of Inorganic Compounds (1904) [in Russian].

[137] V. A. Palauzov, Chemical Reagents; Their Properties, Preparation, Methods of Testing and Use (State Sci. Tech. Press Ukrainian SSR, 1935) [in Russian].

[138] G. Borneman, Inorganic Preparations (State Chem. Tech. Press, 1934) [in Russian].

[139] Chemists' Handbook (State Chem. Press, 1954) [in Russian].

[140] Chemical Reagents and Preparations (State Chem. Press, 1953) [in Russian].

[141] L. G. Berg, A. V. Nikolaev and E. Ya. Rode, Thermography (Acad. Sci. USSR Press, 1944) [in Russian].

[142] A. I. Tsvetkov, Trans. Fourth Conference on Experimental Petrography, No. 1 (Acad. Sci. USSR Press, 1951) [in Russian].

[143] G. N. Voronkov, Trans. Fourth Conference on Experimental Petrography, No. II (Acad. Sci. USSR Press, 1953) [in Russian].

[144] A. V. Nikolaev, Trans. Second Conference on Experimental Petrography (Acad. Sci. USSR Press, 1937) [in Russian].

[145] K. M. Fedot'ev, Trans. Third Conference on Experimental Petrography (Acad. Sci. USSR Press, 1940) [in Russian].

[146] K. M. Fedot'ev, Trudy Inst. Geol. Nauk, Akad. Nauk SSSR [Trans. Inst. Geological Sciences, Acad. Sci. USSR] 120, Petrog. Ser. [Petrographic Series] 35, 1949.

[147] A. I. Tsvetkov, Trudy Inst. Geol. Nauk, Akad. Nauk SSSR [Trans. Inst. Geological Sciences, Acad. Sci. USSR] 106, Petrog. Ser. [Petrographic Series] 30, 1949.

[148] G. G. Tsurinov, The N. S. Kurnakov Pyrometer (Acad. Sci. USSR Press, 1953) [in Russian].

[149] A. I. Tsvetkov et al., Material on the Thermal Investigation of Minerals (Acad. Sci. USSR Press, 1952) [in Russian].

[150] N. I. Gorbunov, I. G. Tsyurupa and E. A. Shurigina, X-ray Pictures, Thermograms, and Dehydration Curves of Materials Encountered in Soils and Clays (Acad. Sci. USSR Press, 1952) [in Russian].

[151] N. A. Shishakov, Problems in the Structure of Silicate Glasses (Acad. Sci. USSR Press, 1954) [in Russian].

[152] B. K. Vainshtein, Structural Electron Diffraction (Acad. Sci. USSR Press, 1956) [in Russian].

[153] K. Kühne, Silikattech. 11, 1956.

[154] B. F. Ormont, Structure of Inorganic Substances (State Tech. Press, 1956) [in Russian].

[155] H. Moore and P. W. McMillan, J. Soc. Glass Technol. XL, 193, 1956.

[156] L. L. Sun and K. H. Sun, Glass Industry 12, 1948.

[157] R. E. Lapp and G. L. Éndryuss, Physics of Nuclear Radiation (Military Press, 1956) [in Russian].

[158] A. N. Winchell, Optical Mineralogy (Foreign Lit. Press, 1949) [Russian translation].

[159] P. P. Solov'ev, Handbook of Mineralogy (Metallurgy Press, 1940) [in Russian].

[160] O. M. Shubnikova, Trudy Inst. Geol. Nauk. Akad. Nauk SSSR [Trans. Inst. Geological Sciences, Acad. Sci. USSR] 74, Mineral. Geokhim. Ser. [Mineralogical — Geochemical Series] 15.

[161] G. Gehlhoff and M. Thomas, Z. tech. Physik. 7, 1926.

[162] P. Gilard and L. Dubrul, Verre et silicates ind. 145, 1934.

[163] P. Gilard and L. Dubrul, Verre et silicates ind. 9 (16) 1938.

[164] A. A. Appen, Zhur. Priklad. Khim. [J. Appl. Chem.] XXIV, 9, 10 and 11, 1951.

[165] L. I. Demkina, Doklady Akad. Nauk SSSR [Proc. Acad. Sci. USSR] VIII, 5, 1947.

[166] M. D. Karkhanavala, Glass Ind. 33, 8, 1952.

[167] Handbook of Physicotechnical Values, Vol. I (State Tech. Press, 1927) [in Russian].

[168] Collection of Physicotechnical Constants (United Sci. Tech. Press, 1937). [in Russian].

[169] Handbook of Chemistry and Physics (Cleveland, Ohio, 1911).

[170] É. Damur, Glass (Ukrainian State Local Industry Press, 1935) [in Russian].

[171] M. A. Matveev and V. A. Kleimenov, Calculations in Glass Technology (State Light Industry Press, 1938) [in Russian].

[172] A. Petzold, Email (Berlin, 1955).

[173] A. A. Appen, Steklo i Keram. [Glass and Ceramics] 1, 1953.

[174] N. J. Ritland, J. Am. Ceram. Soc. 38, 2, 1955.

[175] G. Rindol, Report to the Fourth International Conference on Glass (Bureau of Technical Information, All-Union Glass Fiber Scientific Research Institute, 1957) [in Russian].

[176] É. M. Bonshtedt-Kupletskaya, Determination of the Specific Gravity of Minerals (Acad. Sci. USSR Press, 1951) [in Russian].

[177] J. Mori and K. Eguchi, Report to the Fourth International Congress on Glass (Bureau of Technical Information, All-Union Glass-Fiber Scientific Research Institute, 1957) [in Russian].

[178] G. F. Brewster, J. Am. Ceram. Soc. Cer. Abstr., Aug. 1952.

[179] N. M. Brandt, J. Am. Ceram. Soc. 34, 11, 1951.

[180] N. J. Kreidl and R. A. Wiedel, J. Am. Ceram. Soc. Cer. Abstr., Aug. 1952.

[181] Sawai Ikutaro, Glass Ind., April, 1957.

[182] L. I. Demkina, Optiko-Mekh. Prom. [Opticomechanical Industry] 2, 1957.